Nutritional and Physiological Disorders in Crop Plants

Other Related Books

Advances in Plant Physiology Vol. 1-11	*Hemantaranjan, A.*
Allelopathy in Crop Production	*Narwal, S.S.*
Allelopathy in Soil Sickness	*Narwal, S.S.*
Biodiversity, Conservation, and Systematics	*Prithipalsingh*
Biofertilizers Technology	*Kannaiyan, S.*
Disease Problems in Vegetable Production	*Gupta, S.K.*
Diseases Management in Arid Land Crops	*Lodha, Satish*
Diseases of Horticultural Crops: Nematode Problems and their Management	*Reddy, Parvatha, P.*
Environmental Physiology	*Hemantaranjan, A.*
Fungal Diseases of Horticultural Crops and its their Management	*Parvatha Reddy*
Insect Pests of Horticultural Crops and their Management	*Parvatha Reddy*
Integrated Nematode Management in Horticultural Crops	*Parvatha Reddy*
Integrated Plant Disease Management	*Sharma, R.C.*
Plant Analysis: Research Methods	*Narwal, S.S.*
Plant Breeding (Thoery and Practice)	*Staskopf, N.C.*
Plant Physiology and Plant Molecular Biology in New Millennium	*Hemantaranjan, A.*
Plant Physiology: Research Methods	*Narwal, S.S.*
Research Methods in Plant Science : Allelopathy Vol. 1-5	*Narwal, S.S.*
Seed Borne Diseases - Ecofriendly Management	*Arya, Arun*
Seed Borne Diseases Objectionable in Seed Production and their Management	*Bhale, M.S.*
The Influence of One Plant on Another: Allelopathy	*Molisch, H.*
The Mutual Interaction of Higher Plants- Allelopathy	*Willis, R.J.*

Nutritional and Physiological Disorders in Crop Plants

C. Rajendran
Professor of Crop Physiology (Retd.)
K. Ramamoorthy
Dean and Professor of Seed Science and Technology
S. Juliet Hepziba
Associate Professor of Plant Breeding and Genetics
Agricultural College and Research Institute, TNAU
Madurai - 625 104, TN, India

SCIENTIFIC PUBLISHERS (INDIA)
P.O. Box 91 JODHPUR (Raj.)

Published by
Tanay Sharma
Scientific Publishers (India)
5-A, New Pali Road, P.O. Box 91
Jodhpur - 342 001
Tel.: +91-291-2433323; Fax: +91-291-2613449
E-mail: info@scientificpub.com
www.scientificpub.com

Print, 2023

ISBN: 978-81-7233-606-6

Lasertype set: Rajesh Ojha
Printed in India

PREFACE

In recent years, the concept of nutrient management has undergone many changes. With the tremendous advances made in the understanding of the role of various fertility elements, greater emphasis has now been given to secondary and micro nutrients. Introduction of fertilizer responsive high yielding varieties and hybrids led to major and micro-nutrient deficiencies, resulting in loss of soil fertility. Identification of deficiency symptoms for amelioration involves careful observation of crop behavoiur and a complete understanding of the nutrient elements in the plant system.

Considering the importance and emerging need in this area, a book on **Nutritional and physiological disorders in crop plants** has been prepared in a simple language. We feel this will help to improve the knowledge on diagnosis of deficiencies of mineral elements and their rectification. The main feature of this guide is the point-by-point description of the various visual deficiency symptoms exhibited by the crop plants.

The book has been written primarily for the students of Agriculture and Horticulture and also for those engaged in this profession.

Madurai
3rd June, 2009

C. RAJENDRAN
K. RAMAMOORTHY
S. JULIET HEPZIBA

CONTENTS

1

INTRODUCTION

An outstanding feature of life is the capability of living cells to take up substances from environment and to use them for the synthesis of their own cellular components or as an energy source. The supply and absorption of chemical compounds needed for growth and metabolism may be defined as "nutrition" and the chemical compounds required by an organism are termed as nutrients.

The green plants require essential nutrients exclusively in inorganic forms and this is how they differ from animals, men and a number of microorganisms, which additionally need organic compounds as foodstuffs. An essential nutrient element is the one which is required for the normal life cycle of an organism and whose functions cannot be substituted by any other chemical compound. Besides, the chemical must be directly involved in nutrition as an essential metabolite or the element must be required for the action of an essential enzyme system.

The criteria for essentiality of elements for plant growth have been defined by **Arnon and Stout** in 1939 as under:

A deficiency of the element makes it impossible for the plant to complete vegetative or reproductive growth stages of its life cycle.

The deficiency is specific to the element in question and as such can be prevented or corrected only by supplying that particular nutrient element to the plant, and the nutrient must take part in metabolism as a metabolite or it must be required for action of an essential enzyme system.

It was found later that out of Arnon's all the criteria of essentiality of an element, only the first holds promise and is widely accepted whereas rest is not found true. In many cases, it has been found that rubidium can replace potassium and strontium can replace calcium. Thus it becomes difficult to decide easily the essentiality of potassium, rubidium, calcium and strontium. According to Arnon's theory, sixteen nutrient elements are found essential for plants.

However, the definition of essentiality proposed by **Nicholas** (1963) is more practical. He advanced the terms "functional or metabolic nutrient" to include any mineral element that involves in plant metabolism whether or not its function is specific. Elements chlorine, silicon, sodium and vanadium are also classified as essential according to Nicholas. The list of elements may be further extended to include those elements, which are a part of specific metabolite, but their requirement by plants is yet to be established.

Besides the above mentioned elements, there are few more elements like rubidium, strontium, nickel, chromium, arsenic, etc. which are found to stimulate the growth and have beneficial effects at very low concentrations. These nutrients are called **"beneficial elements"** and they affect the uptake, translocation and utilization of essential elements, help in production of essential metabolite by activating enzymic system/action and also counteract the toxic effects of some other elements or antimetabolites.

The essentiality of nutrient can be studied by raising plants in distilled water or purified sand filled in glass containers made up of hard borosilicate glass viz. pyrex and borosil. The nutrient solution of different elements are prepared and the relative effect is studied by comparing with control (untreated) plants. The trials are not conducted in soil just to avoid various contaminations of different minerals and other elements.

2

CLASSIFICATION AND FUNCTIONS OF PLANT NUTRIENTS

Plant nutrients have been classified in several ways as described below:

1. According to their relative requirements for plant growth

Arnon and Stout (1939) classified the essential plant nutrients based on their relative requirement by the plant:

a. Major, primary or macronutrients

This group includes nitrogen, phosphorus, and potassium, which are required in maximum quantities by the plants. These nutrients are often deficient in almost all the soils because of their heavy depletion. The deficiency is corrected by application of fertilizers based on soil analysis and crop requirement. These elements are also referred as fertilizer elements. This group also includes C, H, and O, which are abundantly present in atmosphere and need not to be applied.

b. Secondary nutrients

This group includes **Ca, Mg** and S. They are required right from the beginning of the plant growth, but in relatively lesser quantities.

c. Micronutrients, trace elements or rare elements

This group includes **Fe, Zn, Cu, Mn, B, Mo** and **Cl**. These elements are required in very minute quantities and they mostly act to stimulate, activate or accelerate the enzymatic activities in metabolic process of plants. These elements are required in traces and hence they are called trace elements. Normally the requirement of these nutrients is met with their quantities present in the soil except few like Zn and S, which need to be supplied separately.

2. According to their biochemical behavior and physiological functions

From the physiological viewpoint, it is difficult to justify the classification of plant nutrients into macro and micronutrients depending upon their concentration in plant tissues. Based on their biochemical and physiological behaviour, the plant nutrients may be classified into following four groups:

Plant nutrient classification

Nutrient elements	Uptake	Biochemical functions
I Group : C,H,O,N,S	As CO_2, HCO_3, H_2O, O_2, NO^-_3, NH_4^+, N_2, SO_4, SO_2	Major constituent of organic matter, essential element of atomic groups which are involved in enzymic processes, assimilation by oxidation reduction reactions.
II Group : P, B, Si	As phosphates, boric acid or borate, silicate from soil solution	Esentrification with native alcohol groups in plants. The phosphate esters are involved in energy transfer reactions.
III Group: K, Na, Mg	As ions from the soil Ca, Mn, Cl solution	Non-specific functions establishing osmotic potentials. More specific reactions by which the conformation of the enzyme protein is brought into optimum status (enzyme activation). Bridging of reaction partners. Balancing diffusible and indiffusible anions.
IV Group: Fe, Cu, Zn, Mo	As ions or chelates	Incorporated in prosthetic groups mostly in chelated form. Enable electron transport by valency change.

Emil Truog and Engelbert (1954) classified the plant nutrients on the **functional basis** along with their usual contents present in plant tissues as a whole. It is mentioned as under:

Group I: Main structural elements of plant tissues

1. **Carbon** constitutes nearly 45 per cent of the total tissues.
2. **Hydrogen** constitutes nearly 6 per cent of the total tissues.
3. **Oxygen** constitutes nearly 43 per cent of the total tissues.

Group II: Accessory structural elements of the more active and vital living tissues, especially proteins

1. **Nitrogen** - 1-3 per cent, constitutes nearly 16-18 per cent of all the proteins.
2. **Phosphorus** - 0.05 to 1.0 per cent. It is constituent of nucleoproteins, phytin and lecithin.
3. **Sulphur** - 0.05 to 1.5 per cent. It is constituent of many proteins and often present as sulphate.

Group III: Regulators and carriers

1. **Potassium** - 0.3 to 6.0 per cent of total tissues. It is not present as constituent but is found as salt and functions mainly in synthesis and translocation of carbohydrates.
2. **Calcium** - 1.0 to 4.0 per cent of total tissues. It is present as calcium pectate and proteinate and as simple organic and inorganic salts.
3. **Magnesium** - 0.05 to 1.0 per cent of total tissues. It forms as essential constituent of chlorophyll and may function as a carrier of phosphorus.

Group IV: Catalyst and activators

1. **Iron** - Present in 10 to 1000 ppm of total tissues. It functions in chloro-

phyll formation and probably in oxidation-reduction reactions. It may also exists in organic combination.

2. **Manganese** - Present in 5 to 500 ppm of total tissues. It is believed to function in chlorophyll formation and also in oxidation-reduction reactions.
3. **Zinc** - It is present in 5 to 100 ppm of total tissues and it functions as catalytic agent
4. **Boron** - It is present in 2 to 75 ppm. Its function in growth processes is not confirmed and probably it functions as a catalyst.
5. **Molybdenum** - It is present in 0.01 to 10 ppm of total tissues. It is required in very minute amount for catalytic processes. Mo rich soils produce forage with about 100 ppm Mo, which becomes toxic to live-stock.
6. **Chlorine** - It is present in 100 ppm to several hundreds of total tissues. Plants need it in highest amount among all the micronutrients.

In the **Handbook of Manures & Fertilizers (ICAR)**, another classification based on functions of elements is mentioned which is as under:

Group I:

Energy exchangers : H,O

Group II:

Energy storers : C, N, P and S.

Group III:

Translocation regulators: K, Na, Ca and Mg.

Group IV:

Oxidation-reduction regulators: Fe, Mn, Mo, Cu, B and Zn.

Functions of plant nutrients

Each of the essential nutrients has a definite and very specific function to perform in the growth and development of plants. The functions performed by individual nutrients are summarized below.

NITROGEN

It is an essential constituent of all the living matter like chlorophyll, protoplasm, protein and nucleic acids. Nitrogen is also found to be present in other compounds of greater physiological importance in plant metabolism like nucleotides, phosphatides, alkaloids, enzymes, harmones, vitamins, etc. Nitrogen imparts dark green color required for an accelerated photosynthetic activities of plants. It increases growth and development of all living tissues specially tiller count in cereals and other graminaceous crops but restricts the root growth. It improves the quality of produce like increased protein content in grains, fodder along with succulence of fodder crops and leafy vegetables. It increases utilization of P and K to an appreciable extent.

PHOSPHORUS

Like nitrogen, phosphorus is also a structural component of the cell constituents and metabolically active compound. It acts as an important structural component of the cell membrane system, the chloroplast and the mitochondria. Phosphorus is a constituent part of the sugar phosphates-ADP, ATP, nucleic acids, nucleoprotein, purine and

Forms of nutrients absorbed by plants and relative mobility of elements in the plant body

Elements	Forms absorbed	Mobility in plants
Nitrogen	NH_4^+, NO_3^- (urea by foliage)	Most mobile
Phosphorus	$H_2PO_4^-, HPO_4^-, PO_4^-$	Most mobile
Potassium	K^+	Mobile
Sulphur	$SO_3^{2-}, SO_4^{2-}, SO_2$	Relatively immobile
Calcium	Ca^{2+}	Relatively mobile
Magnesium	Mg^{2+}	Mobile
Boron	BO_3^{2-}	Relatively immobile
Copper	Cu^{2+}	Relatively immobile
Iron	Fe^{2+}, Fe^{3+}	Relatively immobile
Manganese	Mn^{2+}, Mn^{4+}	Relatively immobile
Molybdenum	MoO_2^{2-}	Moderately mobile
Zinc	Zn^{2+}	Low mobility
Chloride	Cl^-	Mobile
Sodium	Na^+	Mobile
Silicon	$Si(OH)_4$	Not reported

pyrimidine nucleotides, flavin nucleotides and various co-enzymes, viz. NADP, pyridoxyl phosphate and thiamine pyrophosphate.

Phosphorus, in combination with organic acids forms essential constituent of the plant cells like esters, phosphatides and phospholipids. Phosphorus, being phosphoric ester of inositol, is a major constituent of phytin. It plays a vital role in photosynthesis, respiration, energy storage, transfer, cell division, cell elongation and several other processes in the living plants. It promotes early root formation and growth. Phosphorus improves the quality of many fruits, vegetables and grain crops. It is also vital to seed formation by stimulating early flowering leading to early maturity. It improves harvest index of crops. Phosphorus, when applied to legumes, enhances the activity of Rhizobia by increasing nodulation and thereby helps in fixing more atmospheric nitrogen.

POTASSIUM

Potassium maintains the cellular organization by regulating the permeability of cellular membranes and keeping the protoplasm in a proper degree of hydration by stabilizing the emulsions of highly colloidal particles.

Thus it helps in maintaining turgor pressure and eliminates water imbalances in plants. Potassium salts are found to buffer the metabolic actions and used to stabilize various enzyme systems. It acts as catalytic agent in activating a number of enzymes, bonds incorporation of amino acids in proteins, synthesis of peptide bonds and phosphate acetyl transferase. ß-galactosidase requires K for their optimum functioning. It increases resistance in plants against moisture stress, heat and frost and against diseases like:

- Leaf blight and stalk diseases
- Wilt and damping off
- Mold and mildew
- Black spot and stem-end rot
- Wild fire
- Leaf spot and dollar spot

Potassium is important for fruit formation and translocation of heavy metals like iron and, in the improvement of crop quality. It strengthens stalks and stems against lodging and invading organisms like insect-pests.

CALCIUM

Calcium is required for the maintenance of cellular organization by regulating the permeability of cell membranes. It plays an important role in neutralizing the charges on the acidic molecules of phosphoric acid and organic acids like citric acid, malic acid and oxalic acid, which are otherwise structural component of chromosomes, wherein it probably binds the DNA to the protein. That is why Ca deficiency causes chromosomal abnormality. It is also required in mitosis in roots. It stimulates microbial activity and increases symbiotic fixation of atmospheric nitrogen. It increases availability of molybdenum and uptake of other nutrients. It acts as an essential co-factor or an activator of numerous enzymes related with hydrolysis (hydrolytic enzymes or hydrolyses).

MAGNESIUM

Magnesium is a component part of chromosome (bearers of the hereditary characters) and polyribosomes (required for protein synthesis). It is a mineral constituent of chlorophyll. Thus it is actively involved in photo-synthesis. Mg helps in phosphate metabolism, plant respiration and activation of several enzymes, most of which are related with carbohydrate metabolism, phosphate transfer and decarboxylations. Mg is also found to be involved in photosynthesis and organic acid metabolism.

SULPHUR

Sulphur is essential for formation of plant protein because it is part of certain amino acids, which act as building blocks of proteins. It is an important constituent of several other biologically active compounds such as vitamins (thiamine and biotin), lipoic acid, acetyl co-enzyme A, ferredoxin and glutathione. Other biologically active sulphur containing compounds, which play an important role in plant metabolism, include sulphate-3 phosphoadenosine-5 phosphosulphate (an important sulphur donor) and is involved in the synthesis of glucosides in mustard oil. Sulphur promotes nodule formation. Although it is not a constituent of chlrophyll, sulphur is involved in carbohydrate metabolism of the plants by activating a number of enzymes, which participate in dark reaction of photosynthesis.

IRON

Iron acts as catalyst in formation of chlorophyll through formation of chlorophyll precursor, protoporphyrin and acts as an oxygen carrier. It is a constituent of porphyrin compounds cytochromes, haem, non-haem enzymes and other functional metalloproteins, e.g. ferrodoxin and haemoglobin in plants. Cytochromes play an important role in oxidative and photophosphorylations during the process of electron transport and photosynthesis respectively. Similarly ferrodoxin, having very low oxidation-reduction potential, helps in reduction of CO_2, atmospheric N and sulphate. Some micro-organisms like Aspergillus niger contains iron in its black pigment (aspergillin) and yeast consists of iron in its pulcherrine.

Iron, being a structural part and an activator of enzyme, affects enzyme activity. It is bound to the protein in ionic form as metalloflavin and as metalloporphyrin. Iron helps in electron transport coupled with oxidative phosphorylation during respiration. The iron, being a constituent of ferrodoxin cytochromes, is involved in respiration linked active uptake of irons. It, being a constituent of ferredoxin, also plays a key role in nitrogen fixation by diverse groups of micro organisms aerobic and anaerobic bacteria and blue green algae.

MANGANESE

It functions primarily as part of enzyme system, specially of nitrite reductase and hydroxylamine reductase, which are related with nitrogen assimilation. It is found to activate several enzymes like oxidoreductases, hydrolases, lyases, ligases and malic hydrogenase. It plays a vital role in photosynthesis and nitrogen metabolism. It is required to activate numerous enzymes participating in Calvin cycle and because of which the photosynthetic efficiency is accelerated. It is found to accelerate germination and maturity. Mn increases availability of P and Ca to plants. It is also required for proper groupings of chromosomes.

COPPER

Copper is necessary in chlorophyll formation of plants. It also catalyses several plant processes like the respiratory oxygen uptake, the activity of certain enzymes, of which it is not a constituent part of activator and metabolism. The copper containing compounds like plastoquinones and plastocyanins are involved in electron transport from chlorophyll to NADP and from water to chlorophyll during primary reactions of carbon assimilation. It is a constituent part of several enzymes involved in cellular oxidation-reductions like oxydoreductase, urate oxidase, nitrite reductase, cytochrome oxidase and ascorbic oxidase.

ZINC

Zinc is a constituent part of certain enzymes like alcohol dehydrogenase in yeasts and galactose oxidase in moulds.

It is also a constituent part of L-lactase dehydrogenase, triose-phosphate dehydrogenase and carbonic anhydrase in higher plants as well as microorganisms. The metal plays a vital role in photosynthesis and nitrogen metabolism. It helps in regulating auxin concentration in plants. It promotes flower setting and helps in proper development of fruits. It helps in carbohydrate transformation and sulphur metabolism. The metal is found to regulate oxidation and reduction processes in plants.

MOLYBDENUM

Molybdenum is bound to flavin and is required to form nitrate reductase, which is concerned with reduction of nitrate to nitrite and from nitrite to ammonia in plants. It affects the activity of several enzymes like catalase, peroxidase and polyphenol oxidase, though Mo is not a constituent part of these enzymes. It inhibits the functions of acid phosphatase.

It is required in the process of nitrogen fixation by each of the aerobic bacterium (*Azotobacter*), the anaerobic bacterium (*Clostridium*), the blue green algae (*Nostoc, Anabaena*, etc) and the symbiotic bacteria (*Rhizobium* sp). It is found to accelerate the rate of photosynthesis, however its exact role in photosynthesis is yet not known. It also helps in carbohydrate metabolism, formation of middle lamella, promotes nitrogen metabolism and synthesis of plant protein.

BORON

Boron is required for proper development and differentiation of tissues. It helps in cell maturation by regulating the formation and lignification of cell wall. It is found associated with reproductive phase in plants and reduces sterility and any malformation in reproductive organs. It is also involved in carbohydrate metabolism, especially in translocation of photosynthates, the sugars in the form of sugar-borate complexes, which pass rapidly through cell membranes then, do free sugars. It is also involved in nucleic acid metabolism. It inhibits the production and accumulation of phenolic acid and phenolic substances in plants; thus necrosis of tissues is checked. It promotes Ca and K metabolisms and synthesis of pectin, ATP, DNA and RNA.

CHLORINE

It is found to evolve oxygen during primary photosynthetic reactions. It improves turgor pressure and balances cations in cell sap. It is assumed to be a constituent part of anthocyanins.

COBALT

Cobalt has not been proven essential for growth of higher plants but nodulating bacteria (symbiotic or Rhizobia) need it for fixing atmospheric nitrogen in leguminous crops. It is a structural component of vitamin B_{12} (cyanocobalamine) synthesised by many bacteria. It is reported to act as catalytic agent to activate certain enzymes like arginase, lecithinase and oxaloacetic decarboxylase.

SODIUM

Sodium has been found to be essential for only two halophytic species of higher plants namely Atriplex vasicaria and Hologeton glomeratus. It is believed that it helps in maintenance of osmotic relations of the cell which facilitates the plants to resist desiccation when the external concentration of salts is too high. It also influences water uptake and under deficient supply of potash it acts as a substitute of K in plant nutrition. However, its precise role in plant nutrition is yet to be understood.

VANADIUM

Its essentiality is yet not proved, though its beneficial effect on green algae has been established. It is believed to act as a substitute for Mo in nutrition of Azotobacter and Rhizobium bacteria. The growth of certain crops like barley, paddy, maize etc. has been found to be improved by vanadium, although its deficiency does not restrict their growth. It is assumed that it takes part in oxidation and reduction process.

3

DIAGNOSIS OF PLANT DEFICIENCY SYMPTOMS

A deficiency of an element essential for plant growth will result in a decrease in the normal growth of the plant and, will affect the yield of a crop. An essential element will often be sufficiently low so that the deficiency will be manifested in some manner by the plant. Symptoms of nutrient deficiencies have been used for many years to diagnose growth problems.

Sometimes both essential and non-essential elements may be sufficiently in excess so as to be toxic to the plant and to detrimentally affect growth and yield of a crop. When this occurs, the plant will usually express definite toxicity symptoms. Such symptoms can be used to determine the source of the problem, and corrective action can be taken if it is not too late.

The deficiency symptoms can be categorized into five types:

1. **Chlorosis,** which is yellowing, either uniform or interveinal, of plant tissue due to a reduction in the chlorophyll formation processes
2. **Necrosis,** or death of plant tissue
3. Lack of new growth or terminal growth resulting in **rosetting**
4. Accumulation of anthocyanin and an appearance of a **reddish colour,** and
5. **Stunting** or reduced growth with either normal or dark green colour or yellowing

Toxicity Symptoms

Certain essential plant nutrients as well as nonessential elements will be absorbed by plants in excess quantity to be toxic to plant growth. Essential plant nutrients that are seldom, if ever, toxic include N, P, K, Ca, Mg, S, Fe, Zn, Na, and Si. These plant nutrients, if taken up in excess, will often cause imbalances with other nutrients and will result in poor plant growth, *i.e.*, lack of general growth, delayed maturity, and plants that are stunted or spindly.

Essential nutrients that are known to occasionally produce toxicity are Mn, Cu, B, Mo, and Cl. Toxicity symptoms are known for some of these. Manganese causes necrosis and crinkling of upper leaves called crinkle leaf in cotton. Excess Cl causes necrosis along the outer edge of older leaves, a symptom quite similar to K deficiency.

Plants often absorb nonessential

elements. The list that can cause toxicity can be quite long. Several heavy metals are in the group and include lead (Pb), cadmium (Cd), nickel (Ni), aluminum (Al), strontium (Sr), rubidium (Rb), and tungsten (Ti). Symptoms due to a toxicity of these elements are seldom definitive or specific.

There may be many reasons for the occurrence of toxicities. Manganese inhibits the uptake of K and competes with Fe and also inhibits root elongation and damages root cell membranes. An excess of Mo will result in the formation of molybdocatechol complexes in the vacuole, and these apparently inhibit growth processes. These complexes compete with the essential elements that are similar in valency and reactions, thereby disrupting essential metabolic processes.

Use of deficiency symptoms to determine nutrient need

Nutrient deficiency symptoms can be used to determine the nutrient needs of crops, especially when they are specific for a particular nutrient. Yellowing along the midrib and eventual dying of lower leaves is a specific indication of N deficiency, whereas chlorosis and necrosis on the outer edge of lower leaves specifically indicate K deficiency.

As discussed earlier, however, many symptoms can be caused by any one of several nutrients or by other conditions. Yellowing and interveinal chlorosis of young, newer leaves could be due to S deficiency or it could indicate a deficiency of one of the metal micronutrients, Fe, Zn, Mn or Cu. Knowledge of soil pH and general soil conditions may be needed to determine whether the yellowing results from lack of S or one of the micronutrients. Sulfur deficiency symptoms seldom appear on high pH soils, whereas deficiencies of Fe, Zn, Mn, or Cu seldom occur on low-pH soils.

Typical deficiency symptoms can also be caused by many other conditions. Certain herbicides, diseases, and insects can cause chlorosis in plants. Waterlogged or droughty soils and mechanical or wind damage can often create problems that mimic deficiencies.

Deficiency symptoms should be used only as a guideline for determining need. Symptoms plus plant and soil analyses together with a general knowledge of crop needs and the chemistry of the soil should all be used in determining crop nutrient needs.

Determining nutrient needs

In a crop production programme, the most difficult requirement for the grower is to estimate accurately the fertilizer needs of the crop.

Start with visual diagnosis. Compare unusual leaf symptoms to those in this book. Symptoms should be noted as early as they appear in the field; symptoms may be less easily classified later on because they have been modified by other factors.

Second, verify the visual diagnosis by comparing analytical results of leaves with and without deficiency symptoms with critical values reported in the plant analysis table 1 (Annexure).

Comparable leaf samples, with and without symptoms, should be collected for chemical analysis at the same time, since the plants may outgrow the symptoms. For example, in most crops, P deficiency induced by low soil temperature may be overcome by a higher soil temperature. Also, when look-alike symptoms appear, chemical analysis can distinguish leaf scorch caused by drought or windburn from that caused by K and Mg deficiencies. When the symptoms have been identified correctly, the analyses will be in the range that has been reported critical for most crops.

Third, fertilize with the required nutrients, either on a trial basis or over the entire field, leaving an unfertilized area for comparison.

Fourth, confirm by taking leaf samples from the field plots after rainfall or irrigation has been sufficient to ensure that the fertilizer added was actually absorbed by the plants and that the deficiency has been corrected.

Fifth, prevent nutrient deficiencies and crop losses in the current and succeeding crops by following a planned plant analysis programme. A systematic plant analysis program can be used not only to prevent nutrient deficiencies but also to avoid over fertilization. Adding fertilizer as insurance when the soil nutrient supply is already adequate is not only uneconomical but is often unacceptable when it is perceived as a possible pollution source.

Success in the five step programme will be enhanced when the physical and chemical properties of the soil are fully characterized before application of the fertilizer materials required to correct any deficiency.

Pre-disposing factors for occurrence of Nutrient deficiencies

Both soil and atmospheric factors greatly influence the occurrence of deficiency symptoms in crop plants. The list of soil or environmental conditions conducive for certain micronutrient deficiencies are briefly discussed below:

MACRONUTRIENTS

Nitrogen deficiency

1. Low soil organic matter
2. Incomplete decomposition of organic matter
3. Water logging situations
4. High degree of leaching
5. Denitrification by soil microorganism

Phosphorus deficiency

1. Low soil P status
2. Low or high soil pH
3. Application of rock phosphate thus making unavailable to current crop
4. Cold / very dry or very wet conditions.
5. High N / K ratio

Potassium deficiency

1. Low soil K status in highly leached sandy soil.
2. Red lateritic soil
3. High (Mg + Ca) / K ratio in the soil
4. Cold / wet or dry soils

Magnesium deficiency

1. Highly leached acid sandy soils
2. Low soil pH
3. Calcareous soils
4. Imbalance K / Mg and Ca / Mg ratio in the soil

Calcium deficiency

1. Sandy soils with inherent low Ca level
2. Low soil pH

Sulphur deficiency

1. Low organic matter in the soil
2. Low soil pH
3. Highly leached acid soil
4. Cold / wet soils

MICRONUTRIENTS

Iron deficiency

1. Highly calcareous soil
2. High soil organic matter
3. Over limed soil
4. High phosphates in soil
5. Soils rich in manganese
6. Poor soil aeration
7. Bicarbonate rich irrigation water as well as in soil
8. High content of soil organic matter
9. Soils rich in manganese
10. Heavy incidences in susceptible genotypes (eg. sorghum and maize)

Zinc deficiency

1. High doses of nitrogen and phosphorus
2. High calcareous and alkaline / sodic soils
3. Over liming or calcium carbonate rich clay soils
4. Low lying & waterlogged situation
5. Addition of larger quantities of poorly decomposed or undecomposed organic manures / residues
6. Prevalent in acid sandy soils and exposed soils due to poor leveling
7. Most common in compact soil due to poor root growth and penetration
8. Severe cold spells in winter
9. Most common in high rainfall areas
10. Presence of higher contents of N, P, Mg, Fe, Cu and heavy metals like chromium, lead, etc.

Boron deficiency

1. Sandy soils low in organic matter and acidic in reaction.
2. Peat / muck soils
3. Calcareous / alkaline soils
4. Heavy lime application
5. Low soil boron (<0.5 ppm) from acid igneous and fresh water sedimentary rocks
6. Drought / High temperature
7. Moderate to heavy rainfall

Copper deficiency

1. Sands and sandy loam soils
2. Peat, muck and organic matter rich soils
3. Soils with high organic matter
4. Recently reclaimed soils (thus the deficiency is also known as Reclamation Disease)

5. Alkaline soils
6. Low lying waterlogged soils
7. Soils rich in N, P and Zn

Manganese deficiency

1. Organic soils or calcareous mineral soils
2. Alkaline soils
3. Soils rich in iron
4. Presence of organic matter, such as husk, mulches etc.
5. Due to heavy lime application
6. Poorly drained soils
7. Moderate to heavy rainfall
8. Low soil temperature and dry weather
9. In soils rich in Fe, Cu and Zn

Molybdenum deficiency

1. Acid soils.
2. Presence of higher contents of phosphates and oxides of Al and Fe status in the soil

Terminology of nutrient deficiencies

Chlorosis

This refers to general yellowing of the leaf tissue. A very common deficiency symptom, since many nutrients affect the photosynthesis process directly or indirectly.

Discolouration

Some deficiencies lower the amount of photosynthesis and chlorophyll which is produced by the plant. Other coloured compounds can then become dominant. When normal nutrient sinks are not available, the plants can store up excess sugars within other compounds which have distinct colors of red, purple, or sometimes brown. The absence of chlorophyll altogether causes the plant to turn white.

Firing

Yellowing, followed by rapid death of lower leaves, moving up the plant and giving the same appearance as if someone touched the bottom of the plants

Interveinal Chlorosis

Yellowing in between leaf veins, but with the veins themselves remaining green. In grasses, this is called striping.

Necrosis

Severe deficiencies result in death of the entire plant or parts of the plant first affected by the deficiency. The plant tissue browns and dies. This is called necrosis. The tissue which has already died on a still living plant is called necrotic tissue.

Stunting

Many deficiencies result in decreased growth. This can result in shorter height of the affected plants.

Rosetting

Stem of plant does not extend normally so that the youngest leaves grow as a tight bunch near the base of the plant.

Bronzing

Leaves develop a shiny purplish tinge overall.

Mobility of plant nutrients

Plant nutrients which can move from places where they are stored to places where they are needed are called plant mobile. Nitrogen, phosphorus and potassium are always plant mobile nutrients. Deficiencies are noticeable first on older tissue. Plant immobile element deficiencies are noticeable first on younger tissue. Calcium and boron are always plant immobile nutrients. Sulphur, chloride, copper, zinc, manganese, iron and molybdenum are intermediate in plant mobility. Under certain circumstances the intermediate elements are mobile. Mobility in intermediate elements may be linked to the breakdown under low nitrogen conditions of amino acids and proteins in older parts of the plant, and the mobility of these organic compounds to younger parts of the plant in the phloem stream. Under good nitrogen availability, these elements are mostly immobile.

Value of plant nutrient deficiency keys

Use of this plant nutrient deficiency key should be considered first as the first step towards understanding deficiency symptoms in the field, secondly as an educational tool to be used in conjunction with soil testing and plant analysis. Environmental stress such as drought, wet conditions, disease, heat and agrochemical interactions can easily be misinterpreted as deficiency symptoms. Photographs of nutrient deficiencies are useful in diagnosis, but field experience and knowledge of field history based on local experience is the best diagnostic aid

Diagnosing nutrient deficiencey sysmptoms in crops plants

The step in identifying deficiency symptoms is to determine whether the deficiency is the result of a mobile or immobile nutrient based on where the symptom is appearing in the whole plant. Mobile nutrients are nutrients that are able to move out of older leaves to younger plant parts when supplies are inadequate. Mobile nutrients include N, P, K, Cl, Mg, and Mo. Because these nutrients are mobile, visual deficiencies will first occur in the older or lower leaves and effects can be either localized or generalized. In contrast, immobile nutrients (B, Ca, Cu, Fe, Mn, Ni, S, and Zn) are not able to move from one plant part to another quickly and deficiency symptoms will initially occur in the younger or upper leaves and be localized. Zn is a partial exception to this as it is only somewhat immobile in the plant, causing Zn deficiency symptoms to initially appear on middle leaves and then affect both older and younger leaves as the deficiency develops.

The following flowchart provides the key to recognize nutrient deficiency symptoms in plants with actual field photos as examples. However, it must be clearly recognized that considerable amount of yield is already lost by the time nutrient deficiencies become apparent in the plants. So to maximize yield and profit soil test and yield

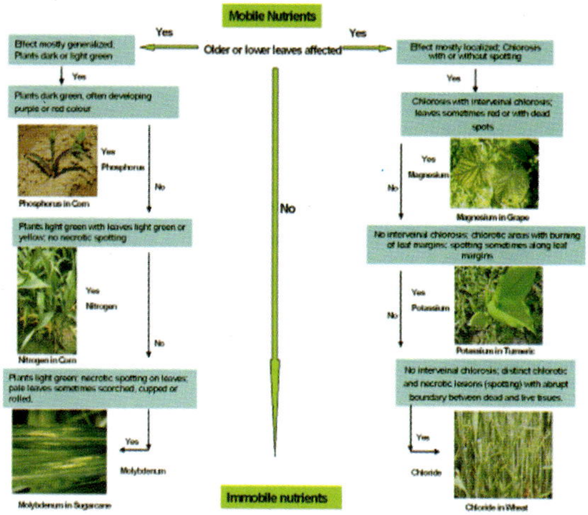

target-based application of nutrients is essential.

Signs of faulty nutrition in plants

When plants are grown in unsuitable environments, including conditions of faulty mineral nutrition, they react to the particular defects in more or less specific ways. Thus, if light is insufficient, the green coloring matter of the leaf will be lacking and the leaves may be almost white (chlorotic), and the plant may be very spindly in appearance; if the temperature is too high, the growth may be rank and soft; if water is insufficient, growth may be restricted, the tissues woody and the green of the leaves show a bluish tint. Again, deficiencies and excesses of the individual elements produce characteristic effects on various organs of plants: foliage characters, including color, density, size and shape of leaves; stem characters, such as thickness, color and length of inter-nodes; root characters, such as color, amount of fiber, abnormal thickening; blossom characters, including amount and time of opening of the flowers; fruit characters, such as size, color, hardness and flavor.

Ability to recognize these particular effects forms the basis of the visual method of diagnosing plant deficiencies. Many of them can be readily learned and applied by farmers. Indeed, for many years progressive fruit growers in this country have used the leaf symptoms of deficiencies of nitrogen and potassium, and more lately of magnesium, as the main guide to manuring their fruit trees and bushes with these elements.

Key to Plant Nutrient Deficiency Symptoms

The basic key points in diagnosing various nutrient deficiencies are enumerated in the following table

Common symptoms of nutrient deficiency in plants

Nitrogen	Entire plant turns yellow green, and the older leaves become more yellowish than the younger. Older leaves do not die unless deficiency is extreme (almost never seen in aquaria)
Phosphorus	Plant stops growing and becomes darker green or stays green. Some species may become purple with excess anthocyanin pigments building up. Other species do not produce excess anthocyanins and just stay green and small.
Calcium	1. Mild deficiency---Smaller, distorted new leaf growth. Reduced leaf tissue, with the central vein persisting. Leaves often cupped, rather than flat. 2. Moderate deficiency---Often sudden bends or twisting of leaf, which is now much reduced in size. White streaks or white edges in new growth. Roots are stubby and twisted. Root tips may di e. Leaves of Vallisneria are strongly crinkled asthough they have tried to grow and got jammed in a small space. 3. Severe deficiency---New growth almost entirely white. Leaves are tiny deformed stumps. Growing points for both shoot and root die.
Magnesium	In dicots---Yellowing of older leaves that starts from the egdes inwards. The midrib may remain green while the edges are yellowed or whitish and dying
Potassium	Small dead areas appear in older leaves. These can start like little pinpoints and grow. In some species, like Ceratopteris, the older leaves stay green while the little dead spots grow. The new leaves are reduced in size and leaf area, looking a bit 'singed'. In other species the older leaves can turn yellow before they die, but they do not have green persisting along the major veins as in magnesium deficiency.
Sulphur	Similar to nitrogen deficiency
Iron	Reduced chlorophyll in new growth. Leaves and stem are about the same shade. Growing tips of Ceratophyllum become pinkish and then white. *Eigeria densa* tips become greenish yellow to yellow with the leaves small and clasped close to the stem. The new leaves of swords are smaller with patches or broad streaks extending lengthwise that are more pale than the rest of the leaf (in mild deficiency). In more severe deficiency in most plants chlorophyll is lacking completely in the new growth which soon dies.
Manganese	Dead yellowish tissue between leaf nerves
Copper	Dead leaf tips and withered edges
Zinc	Yellowish areas between nerves, starting at leaf tip and edges
Boron	Very similar to calcium deficiency. New growth is distorted and smaller, and then the growing tips of both roots and shoots die. In mild deficiency, the leaves are cupped and the roots are shorter and distorted
Molybdenum	Yellow spots between leaf nerves, then brownish areas along edges. Inhibited flowering

The most preferred way to tackle nutrient deficiencies is to determine annual crop's nutritional needs with soil testing, perennial crop needs with plant analysis, and to correct the deficiencies before the crop is established or deficiency symptoms appear. However, soils are very variable in nutrient levels. Parcels of fields low in a certain nutrient may not show up in a normal composite soil sample. The chemistry of some nutrients, such as iron and manganese, is very complex and difficult to predict with a simple soil test. Thus, deficiencies of some nutrients can sometimes appear even with a good soil testing programme. Certain nutrient problems are sometimes so rare that regular soil analysis requests do not include testing for those nutrients. Nutrient deficiency symptoms are often the first clues to a nutrient problem within a field.

Plants which are under stress due to nutrient deficiency show unusual growth patterns or colouration. These visual symptoms are called deficiency symptoms. Deficiency symptoms can sometimes be confused with other complex field events, such as high water tables, salt damage, disease, drought, herbicide stress and varietal differences. Deficiencies can also be so slight that they are confused with other problems. If more than one deficiency is present, one can be more dominant in its symptoms, obscuring the symptoms of the other element.

Deficient element and deficiency symptoms

A. CEREALS

1. RICE

Nitrogen deficiency

- ❖ Stunted growth
- ❖ Reduction in tillering and panicle size
- ❖ Reduction in grain weight
- ❖ Leaf chlorosis followed by necrosis, first in old leaves and then in others
- ❖ A slight delay in heading

Phosphorus deficiency

- ➢ Severe stunting at seedling
- ➢ Small, erect and dark green leaves
- ➢ Reduction in stem thickness
- ➢ Lack of tillering
- ➢ Delayed plant development and maturity

Potassium deficiency

- ❖ Stunted growth, slight reduction in tillering
- ❖ Short droopy and dark green upper leaves
- ❖ Lower leaves exhibit yellowing of interveinal areas starting from leaf tip and eventually drying as indicated by light brown colour
- ❖ Presence of occasional brown spots on upper leaves

Sulphur deficiency

- ➢ Upper leaves display chlorosis similar to N deficiency (as manifested in lower leaves)

➢ Unlike in the case of N, lower leaves do not become necrotic prematurely with S deficiency

Iron deficiency

❖ Entire leaf becomes chlorotic and then bleached to white

❖ Symptoms exhibit first on the young and emerging leaves

Zinc deficiency

➢ Deficiency symptoms exhibited variously like basal chlorosis of leaves, loss of turgidity, delayed plant development; Bronzing and necrosis of leaves and in some cases, death of seedling

➢ In direct dry seeding cultural system, symptoms often occur shortly after flooding, intensified in deep water

➢ Initial symptoms are followed by necrosis of entire plant, especially if the flood water is not removed.

➢ Maturity delayed by 3 - 4 weeks.

Manganese deficiency

❖ Stunting of plant and interveinal chlorosis of new leaves (symptoms likely to appear under high pit soils under nonflooded conditions)

Silicon deficiency

➢ Leaves soft and droopy

➢ Increased lodging; Presence of brown spots resulting from infestation by Helminthosporium

➢ Reduced grain yield

2. SORGHUM

Nitrogen deficiency

❖ Plants stunted, spindly with pale green to pale yellow coloured leaves

- ❖ Pale to deep yellow colour appears near the tip and margin and advance towards the base and midrib of leaf
- ❖ Under severe deficiency, dark brown necrotic spots develop; leaves turn pale brown, die and bent over or fall down (pendent) on the plant
- ❖ Growth of plant very slow, flowering delayed, leaf size reduced
- ❖ Reduced yield with fewer grains

Phosphorus deficiency

- ➢ Common during cool weather in young plants, young plants more erect and sometimes leathery
- ➢ Shoot growth more affected than roots, hence shoot root ratio decrease
- ➢ Grain development and filling are inhibited leading to shriveled kernels of poor quality
- ➢ Plants stunted and spindly with low vigour and dark green leaves which have overtones of dark red colouration
- ➢ Symptoms first appear on older leaves and sheaths. Leaf tips and interveinal tissue show red coloration that progresses towards base and midrib
- ➢ Eventually the entire leaf affected
- ➢ Roots turn dark brown, purple or black

Potassium deficiency

- ❖ Symptoms usually appear in older leaves but in some varieties, middle leaves first affected
- ❖ Irregular necrotic patterns intermingled with red or brown pigmentation
- ❖ Symptoms begin at leaf tips and margins and move towards the base and midrib
- ❖ Necrotic interveinal lesions appear near the margin and sometimes extend for most of the leaf length
- ❖ Reduced grain yields

Calcium deficiency

- ➢ Severe deficiency causes plants stunted since inter nodes fail to elongate and new growth grows in a rosette form
- ➢ Symptoms appear first in newly emerging leaves
- ➢ Serrated upper leaves whose margin show irregular bleached or pale green streaking
- ➢ Leaves short, erect and easily torn

- ➢ Leaf tips fail to unfold, and deformed and they form sword like projection
- ➢ These tips sometimes envelop other leaves to form a ladder like appearance with blade tips sticking together
- ➢ Leaves brittle turn brown and form sticky residues near margin since apical meristems destroyed
- ➢ More tillering

Magnesium deficiency

- ❖ Plants stunted and less vigorous
- ❖ Symptoms first appear on older leaves and then on younger leaves
- ❖ Larger irregular necrotic spots or lesions on leaf tips and margins which progress towards base and midrib

Sulphur deficiency

- ➢ First upper leaves turn pale yellow and emerging leaves uniformly pale yellow
- ➢ Prolonged deficiency makes older leaves also pale yellow with chlorosis developing first near the tip and advancing along the margin towards the base
- ➢ Chlorotic tissue turns brown and leaves die and hang down

Manganese deficiency

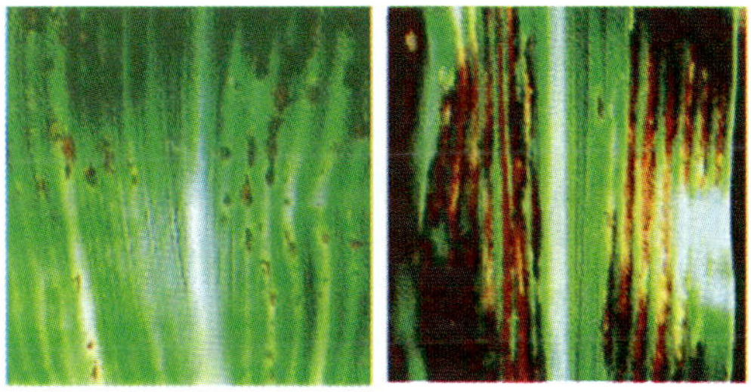

- ❖ Initially middle to upper leaves show interveinal lesions near the middle portions, which progress towards the tip.
- ❖ Long chlorotic streaks along with smaller dark brown or red lesions which form a long streak of red / dark region
- ❖ Under mild deficiency veins remain green with long interveinal lesions

Iron deficiency

- ➢ Symptoms appear first on young leaves with long yellow interveinal streaks, leaf tips and margin more yellow than midrib or base

- When deficiency is severe, leaves more yellow with bleached white appearance and leaf tips and margin turning brown and dying, reduced growth and yield with delayed flowering and maturity

Zinc deficiency

- Symptoms appear first on young leaves, leaves near the whorl show elongated, pale green to bleached white sections
- Plant stunted with shorter inter nodes
- Some genotypes may have reddish purple pigmentation on the under-side of affected leaves, or red brown sections next to or over the bleached portion
- Sometimes, a narrow red band appears along the margin
- Flowering and maturity delayed

Copper deficiency

- First leaves within the whorl and young mature leaves turn pale yellow particularly in interveinal tissue.
- Emerging leaves remain tightly curled
- Pale brown necrosis develops near the tip of younger leaves; leaf tip and margins wither and die to form a twisted spiral; twisted brown tip hangs down from green basal tissue of leaf
- Growth stunted and depressed

Boron deficiency

- Leaves dark green, short, erect, brittle and easily torn or broken.
- Leaves emerging from the whorl develop interveinal chlorosis in the middle of the leaf
- Severe deficiency destroys apical meristems and tillers develop, if it is before boot stage, tillers may appear from nodes well above the crown

3. MAIZE

Nitrogen deficiency

- In young plants, stunted and spindly growth and yellowish-green foliage
- Later, yellowing is predominant
- Tips of the lower leaves are first to show yellowing
- Yellowing will follow up the leaf mid -ribs in a typical "V" shaped pattern with leaf margins remaining green

Phosphorus deficiency

- Stunted growth
- Severe deficiencies cause purpling or browning of leaves; starting with older leaves and moving up the plant and from the tips of the leaves inward
- Leaf tips start to die and turn dark brown
- Delayed maturity; noticeable during pollination; resulting in imperfect ear formation, irregular kernel rows, a twisted lock of the ear and imperfectly developed ear tips

Potassium deficiency

- Leaves are light green or streaked with yellow stripes
- Tips and margins of the leaves become necrotic

- In older plants similar symptoms; marginal browning of the leaves is more conspicuous
- Leaves show scorched appearance; green middle portion exhibit ragged appearance
- By prolonged deficiency, plants become dwarf and weak
- Though stunted, the plants tend to lodge because of the poorly developed root system and susceptibility to stalk rot
- Internodes are shortened
- Corn ears are small, tips are pointed; grains are not at all formed at the tip portion (called tip sterility). Grains are also of poor quality.

Remarks

- Foliar spraying with muriate of potash at 1 to 2 per cent

Magnesium deficiency

- Observed in acidic sandy soils, In seedlings, symptoms appear on the lower leaves as general yellowing and eventually the areas between the veins become light yellow to almost white
- Definite striping seen.

- Under severe deficiencies, tips and the edges of the lower leaves may die and show interveinal striping

Sulphur deficiency

- Stunted growth, delayed maturity, general yellowing of the foliage of interveinal pattern appears
- Yellowing usually more marked in younger leaves

Calcium deficiency

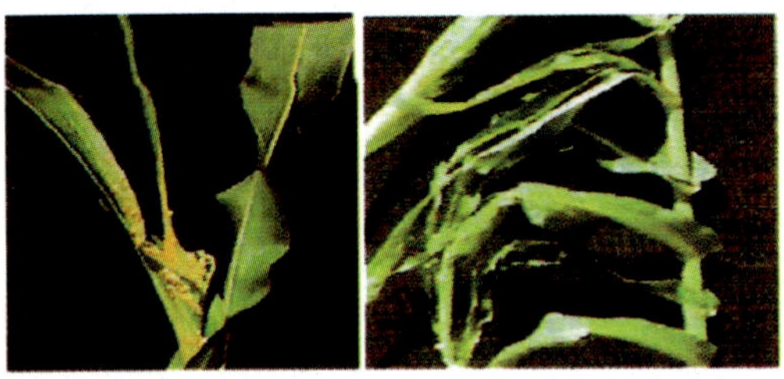

- Inability of the leaves to unfold or emerge
- Tips of leaf stick together, ladder like appearance of plants
- Severely stunted. In extreme cases, the growing tips of the plants die

Iron deficiency

- Most common in calcareous soils
- Interveinal chlorosis or striping which extends over the entire length of the leaves
- Chlorosis most severe on the upper leaves
- In severe iron deficiency, plants become white and eventually die

Zinc deficiency

- Characterized by a broad band of bleached tissue of each side of the mid rib, beginning at the base of the leaf
- Symptoms occur mainly in lower half of the leaf

Boron deficiency

- ❖ Youngest leaves appear white, irregularly shaped spots scattered between the veins
- ❖ With severe deficiency, the spots may form white stripes one to two inches long
- ❖ White stripes appear to be waxy and raised from the leaf tissue
- ❖ Shortening of internodes and often the young leaves fail to unfurl

Copper deficiency

- ➢ Yellowing of the upper or youngest leaves and slight stunting in growth
- ➢ Areas of dead tissue appear along the tip and edges of the leaves
- ➢ Younger leaves become quite pale yellow, and die back of the older leaves occur
- ➢ Plants show wilted appearance

Molybdenum deficiency

- ❖ Older leaves die at the tips, then along the margins necrosis develops between the veins, and eventually die

4. WHEAT

Nitrogen deficiency

- ➢ Pale green leaf colouration
- ➢ Lower leaves show symptoms first
- ➢ Plants are stunted

Phosphorus deficiency

- ❖ Reduced tillering
- ❖ Severe shortages leads to leaf tip die back

Potassium deficiency

❖ Chlorosis appear uniformly at first on older leaves

❖ Leaves become streaked with yellow stripes

Copper deficiency

➢ Leaves light green and become dry at tips

➢ Leaves shriveled, twisted and broken

➢ Stunted growth

Magnesium deficiency

❖ Interveinal chlorosis on the older leaves

❖ Older leaves become reddish purple and the tips and margins may die

Manganese deficiency

➢ Plants show chlorotic leaves

➢ Greenish gray spots develop on the older leaves

➢ Severely affected leaves may droop

Sulphur deficiency

❖ Yellowing and stunting of the plant

❖ Delayed maturity

Zinc deficiency

- Interveinal chlorosis on young leaves
- Stunted plants
- Under severe condition, leaves may turn white and die

B. PULSES

1. RED GRAM

Nitrogen deficiency

- Leaves pale green, greenish yellow or pale yellow
- When deficiency is severe, older leaves show interveinal chlorosis

Zinc deficiency

- Severe deficiency produces narrow, tapered and chlorotic leaves
- Mild deficiency has the effect of nitrogen shortage
- Bunches small and misshapen with small, twisted, chlorotic bottle necked fruits
- Young leaves bunched or chocked

Phosphorous deficiency

- Plants small and stunted
- Leaves dark green

Potassium deficiency

- Leaf tip yellow or brown; yellowing spreads from tip along the margin and may coalesce

Magnesium deficiency

- Mild interveinal chlorosis; veins remain dark green.
- Interveinal areas become rusty brown, or bronzed and necrotic
- Margins of young leaves rolled inward

2. BROAD BEAN

Phosphorus deficiency

- Stems thin, leaves erect early defoliation of basal leaves.

Calcium deficiency

- ❖ Young growth deformed, young stems and petioles wilt and growing points die
- ❖ Leaflets near growing points fail to expand, distorted and hooked, and tips die.
- ❖ Pods deformed, wilted and blackened; seeds fail to develop.

Magnesium deficiency

- ➢ Central interveinal chlorosis, margins green.

Potassium deficiency

- ❖ Internodes short; leaves dark brown marginal scorch
- ❖ Margins curled forward.

Manganese deficiency

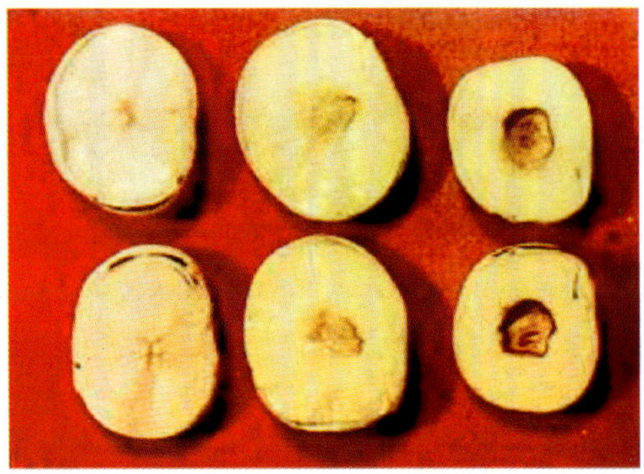

- ➢ Brown lesions in centers of cotyledons, similar "Marsh Spot" in peas

Boron deficiency

- ❖ Stems somewhat stiff; growing points die and lateral growth develop from bases of shoots
- ❖ Leaves slight interveinal chlorosis.

3. FRENCH BEAN

Nitrogen deficiency

- Growth dwarfed and thin
- Leaves pale green and older leaves yellow and die early.

Phosphorus deficiency

- Growth dwarfed and thin: leaves dark, lustreless green
- Older leaves turn brown and shed early.

Calcium deficiency

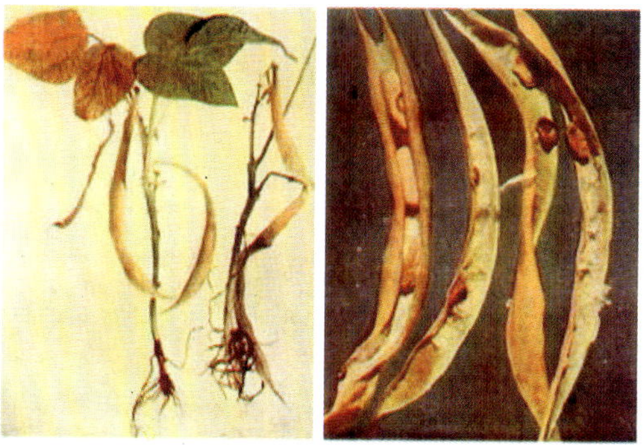

- Growth dwarfed and thin
- Stems, petioles, pedicels and pods wilted
- Early defoliation
- Pods wilted and deformed; seeds fail to develop.

Magnesium deficiency

- Older leaves chlorotic mottling, followed by general yellowing and brown necrotic spotting.

Potassium deficiency

- Interveinal chlorosis near margins followed by marginal scorch.

Manganese deficiency

- Leaves severe chlorosis and necrosis

Boron deficiency

- ➢ Stems abnormally thickened; longitudinal splitting of epicotyl;
- ➢ Death of growing point;
- ➢ Leaves show chlorosis

4. PEAS

Phosphorus deficiency

- ❖ Growth dwarfed and thin; leaves dull, lustreless green;
- ❖ Older leaves wither and die early.

Potassium deficiency

- ➢ Young internodes very short and plants squat;
- ➢ Older leaves show marginal scorching

Calcium deficiency

- ❖ Young stems, pedicels and leaf tissues wilt and collapse.

Magnesium deficiency

- ➢ Central interveinal chlorosis and green marginal bands.

Manganese deficiency

- ❖ Interveinal chlorosis beginning at margins
- ❖ Brown lesions in centers of cotyledons. ("Marsh Spot")

Boron deficiency

- ➢ Stems thickened and stiff, growth squat and bushy habit;
- ➢ Foliage chlorotic, young leaflets small and tips brown
- ➢ Growing points die.

5. GREEN GRAM

Manganese deficiency

- ❖ Plant growth stunted; fewer pods and auxillary branches; smaller leaves
- ❖ Flowering and pod formation inhibited, even when formed they are shed prematurely
- ❖ Middle leaves show fading of leaves followed by interveinal chlorosis
- ❖ Leaves chlorotic, then develop grayish brown spots which turn dark brown necrotic spots particularly along the apical margins
- ❖ Middle leaves exhibit fading of lamina
- ❖ Leaves become severely chlorotic and, develop minute grayish brown
- ❖ Spots which turn dark brown become necrotic
- ❖ Plants short and possess very few branches with small leaves
- ❖ Flower number reduced and shed prematurely

Zinc deficiency

- ➢ Suppression of branching, reduced leaf expansion, delayed and poor flowering and failure to set fruits
- ➢ Apical leaf margins of second trifoliate leaf curl inwards and interveinal leaf areas turn light green and show brown pigmentation, which is prominent on the lower surface of leaf;
- ➢ Eventually, dark brown pigmentation spreads to entire lamina (except margins) which turns necrotic, brittle and crumpled.
- ➢ Leaf apices and margins appear dry, scorched and curled upwards
- ➢ Leaf veins thick (corky) and undulated, causing puckering of leaflets
- ➢ When there is appearance of third set of trifoliate leaves, onward curling of leaf spots develop in the interveinal leaf areas
- ➢ Spreading of dark brown spots from the basal part of the lamina towards leaf apices and margins; then dark brown areas turn necrotic
- ➢ Veins close to the base of leaflets exhibit reddish pigmentation particularly on the under surface of the leaflets
- ➢ Branching, heights, number of expanded leaves, leaf area, flower-

ing and settling of fruits all restricted

Iron deficiency

- ❖ Chlorosis of young leaves and tissue necrosis
- ❖ Development of axillary buds and necrosis and scorching of green tissues

Boron deficiency

- ➢ Distortion, drying and death of apical growths even before the development of trifoliate leaves
- ➢ Under severe deficiency, even the cotyledons may show yellow green mottling and marginal scorching

6. SOYA BEAN

Nitrogen deficiency

- ❖ Initially, the plant green later the leaves turn uniformly yellow.
- ❖ Basal leaves show symptoms first which spread to upper parts, plants eventually defoliate and are often spindly and stunted.

Remarks

- ➢ N deficiency symptoms are occasionally caused by soybean cyst nematodes, root rot or Mo deficiency

Phosphorus deficiency

- ❖ Growth retarded, plants spindly and have small leaflet
- ❖ Leaves turn dark green or brownish green.
- ❖ Leaf blade curl up and appear pointed
- ❖ Flowering and maturity delayed

Remarks

- ➢ `P' uptake is reduced in cool, wet soils

Calcium deficiency

- ❖ Emergency of primary leaves delayed, and when emerged, they are cup shaped

- ❖ Terminal buds of primary leaves become necrotic and other parts chlorotic; Tissues between veins tend to be ridged
- ❖ Terminal buds wither and petioles fall
- ❖ Primary leaves become flaccid and drop prematurely
- ❖ Poor nodulation

Potassium deficiency

- ➢ Deficient plants have weak stems and are more susceptible to diseases.
- ➢ In early stages, an irregular yellow mottling appears around leaflet margin, the yellow areas coalesce to form a more or less continuous irregular yellow border; this is followed by necrosis of chlorotic areas and downward cupping of leaf margins, the dead tissues drop away so that the leaves appear ragged. Chlorosis and necrosis may spread inward to include half or more of the leaflet.
- ➢ Severe deficiency tends to produce misshapen and wrinkled seeds

Magnesium deficiency

- ❖ In early stages of growth, pale green to yellow interveinal leaf and later, turn deep yellow except at the base.
- ❖ Symptoms first appear on lower leaves, later rusty specks and necrotic blotches between the veins and around the edges of middle and upper leaves
- ❖ Leaf margins curl down show yellowing followed by bronzing of entire leaf

Sulphur deficiency

- ➢ Top leaves small, yellow, green, stems thin, hard
- ➢ Elongation of stem is a characteristic symptom

Iron deficiency

- ❖ Yellowing of interveinal areas of younger leaves form early symptoms; later, veins chlorotic and finally entire leaf turns ivory to white in colour
- ❖ Leaf margins show brown necrotic spots

Remarks

- ➢ Deficiency severe in high pH soils and poorly aerated soils with high contents of Mn

Zinc deficiency

- ❖ Stems stunted and young leaves with interveinal chlorosis
- ❖ Later entire leaves turn brown or grey and drop early; few flower and abnormal pods which slow in maturity.

Manganese deficiency

- ➢ Interveinal areas of leaves light green to white and veins green
- ➢ Symptoms resemble early symptoms of Fe deficiency and difference is that leaves remain green and stand out prominently
- ➢ When severe, necrotic brown spots develop

Remarks

- ❖ Symptoms more pronounced in cool weather and under conditions of high Fe concentration

Boron deficiency

- ➢ Shortened internodes and yellowing or reddening of upper leaves
- ➢ Flower fail to develop in dry weather

Molybdenum deficiency

- ❖ Leaves pale green or yellow, necrotic and twisted

- ❖ Margins, midribs and interveinal areas necrotic
- ❖ Yield greatly reduced with fewer pods and seeds

Remarks

- ➤ Deficiency occurs in highly acidic soils.

C. OILSEEDS

1. GROUND NUT

Nitrogen deficiency

- ❖ Young plants appear lighter green and in severe cases, the entire leaf becomes uniform, pale yellow stems slender and elongated
- ❖ Lower older leaves most affected, stems appear reddish because of accumulation of anthocyanin pigments

Remarks

- ➤ Lack of nodulation due to inadequate amount of soil bacteria (or) insufficient N reduction attributable to Mo deficiency (Under extreme soil acidity)

Magnesium deficiency

- ❖ Interveinal chlorosis first in older leaves and then in younger ones
- ❖ Yellowing begins at leaflet margin and extends towards midrib
- ❖ Leaflet margin turns orange and crinkle or curl
- ❖ Older leaflets develop necrotic spots and defoliate

Remarks

- ➤ Magnesium may be added in the form of dolomite, limestone. This will correct the acidity and supply both Mg and Ca

Phosphorus deficiency

- ❖ Leaves first appear bluish green; as leaves mature, they become thickened and leathery characterised by dark green colour, then older leaves turn orange yellow & veins become reddish brown, stem also affected
- ❖ Leaf size markedly reduced and plant maturity delayed
- ❖ Reduced flower production and fertilization
- ❖ Pod yields reduced

Remarks

- ➤ Besides N, P is often the most limiting element

Potassium deficiency

- ❖ Interveinal chlorosis or yellowing first appear at the leaf margins, which change to reddish brown and then become necrotic or scorched
- ❖ In severe cases, affected older leaves are shed, leaf margins curl upward, tips of branches redden and die

Sulphur deficiency

- ➢ Root development restricted, new leaflets pale green or yellow (chlorosis)
- ➢ Plants stunted with reduced branching and hence, appear upright

Remarks

- ❖ S is also absorbed through stomata as SO2 released from burning coal and wood on reaction with water in the cells; SO2 is converted to bisulfite which inhibits photosynthesis and causes chlorophyll destruction. SO2 injury to plants is caused by acidic aerosols during periods of foggy weather, light rains, high RH and moderate temperatures.

Iron deficiency

- ➢ Young leaflets become pale green with veins remaining green. In severe deficiency, leaves become almost white; affected leaves develop brown spots or necrosis on the lamina. Plant growth restricted and leaf size reduced

Remarks: Zn toxicity

- ❖ Plants chlorotic; stunted during seedling stage; progressive necrosis and finally necrosis of entire plant. Purple colouration of main stem and petioles. In severe cases, lesion develops on the stem near the base of the plant and hence, premature death of plant.

Manganese deficiency

- Interveinal chlorosis of young leaflet; may appear bronzing symptoms
- Older leaves necrotic and fall, stems slender and weak. Significant reduction in fruiting and pod yield.

Copper deficiency

- Young leaves deformed and greenish yellow or chlorotic; Terminal leaflets small; margins curl inward giving a cupped appearance
- White spots occur in affected part
- Tips and margins develop necrotic spots
- Bud areas affected leading to stunted growth with short branches
- Reduced pod yield

Boron deficiency

- Leaves deep green; water soaked areas and lesions on leaves, petioles and stems
- Restricted plant growth
- Terminal leaflets small & deformed.
- Internodal length reduced and secondary branching occurs, hence short & stumpy plant and lower branches split
- Reduced flowering & pod production
- Root system stunted, brown and brittle
- Root tips decayed, lateral roots shorter and thicker; seeds poorly developed, innerface of the cotyledons depressed.

2. GINGELLY

Nitrogen deficiency

- Plants light green
- Stalk slender
- Branches absent and leaves erect
- Lower leaves become lemon yellow turning to orange colour
- Discoloured leaves are shed

Phosphorus deficiency

- Branches suppressed
- Stalks slender
- Lower leaves dull dark grayish-green
- Necrosis of lower or majority of leaves and then defoliation.

Magnesium deficiency

- Lower leaves develop interveinal chlorosis
- Light yellow in colour, later becoming orange
- Green colour persists in midrib and veins giving a characteristic pattern
- Leaves develop interveinal chlor-

osis and at later stage, chlorotic tissues develop light brown or husk coloured necrotic lesions

Potassium deficiency

- Young plants have dark green leaves with small stems and shorter internodes
- Plant tips have a spike like appearance because of slower development of leaf petioles
- Young leaves dark green, crinkled and curled
- Older leaves chlorotic and bronzed, leaf margins brown and tissue breakdown

Zinc deficiency

- Young leaves smaller with yellow interveinal mottling
- Older leaves show necrotic interveinal areas

Manganese deficiency

- Young leaves show interveinal chlorosis with prominent green venation
- Under severe deficiency, necrosis of interveinal tissues occurs

Boron deficiency

- Growing tip affected; terminal shoot turns inward and dies
- Leaves small, crinkled, deformed with large irregular areas of discolouration

Molybdenum deficiency

- Leaves show a characteristic cupping or curling in the margins; yellowing between veins and death of growing tip

3. COCONUT

Nitrogen deficiency

- Yellowing of leaves and stunted growth
- Reduction in yield

Phosphorous deficiency

- Retarded root growth

- Delayed flowering and ripening of nuts

Potassium deficiency

- Leaves are short, chlorotic and scorched; crowns are poorly developed.
- First symptoms appear on leaflets of older leaves as scattered pale green spots; with age, these spots may enlarge and turn orange yellow and then reddish brown
- Leaf margins dry starting from the tip
- Palms deficient in K take a longer time to come to bearing and produce few small nuts

Magnesium deficiency

- Leaflets show yellow discolouration while the base remains green

Sulphur deficiency

- Typical symptoms are yellowish green or yellowish orange leaflets
- In severe deficiency, there is complete chlorosis and necrosis of the leaflets
- Leaves drop as the stem becomes weak
- In older palms, leaf size and number are reduced and sometimes an apron of dead fronds develops around the stem due to weakness of the rachis. Nuts may fall pre maturely.
- Copra from S deficient palms is rubbery and unmarketable.

Zinc deficiency

- Dwarfing and deformed growth of tender leaves called "rosette" or little leaf formation

Boron deficiency

- Most common symptoms are fusion of terminal pinnae or 'hook' or 'bayonet leaf' in which the pinnae are bent into double or single hook near the tip. In most severe cases, fronds may develop without pinnae
- Early shedding of buttons (female flowers), resulting in poor fruit set,
- Small and unevenly opened leaves.

D. FIBRES

COTTON

Nitrogen deficiency

- If N is deficient during initial and mid season, older leaves turn yellow to green and then other leaves affected; leaf size reduced

- Plant height reduced, few vegetative branches, fruiting branches short.
- Boll shedding during 10-12 days after flowering
- When N deficiency occurs in late season, reddening of leaf in the middle of canopy and few bolls only retained
- Early maturity of bolls due to accelerated leaf senescence & shorter flowering period

Phosphorus deficiency

- Plants stunted, leaves dark green
- Flowering delayed; boll retention poor
- Premature senescence

Potassium deficiency

- Older leaves first show interveinal yellowing, necrotic patches at leaf margins under severe deficiency

- After peak bloom, symptoms first appear on young mature leaves in upper third of the canopy
- First as slight interveinal yellowing which change to bronze orange colouration
- Leaves curl downward & become thickened and necrotic patches at margins,
- Boll retention poor and premature defoliation

Calcium deficiency

- Root growth affected, root tips brownish, lateral branches reduced
- Under severe deficiency, yellowing and necrotic lesions on cotyledons
- Terminal shoot meristem dies

Magnesium deficiency

- Older leaves first show a distinct reddish purple interveinal discolouration
- In severe cases, other leaves also affected

Sulphur deficiency

- Younger leaves in the upper canopy first show pale green colour and then yellowish similar to N deficiency but leaf veins remain greener
- Plants short, have few vegetative branches and small bolls

Manganese deficiency

- Younger leaves show interveinal chlorosis and leaf cupping
- Delayed appearance of first flower

Zinc deficiency

- Interveinal yellowing with distinct green veins on younger leaves; smaller leaves, leaves thicken and abscise

Remarks

- Deficiency common in calcareous soils and soils with high P content.

Iron deficiency

- Younger leaves show interveinal yellowing and veins remain green.

E. SUGARS

1. SUGARCANE

Nitrogen deficiency

- Older leaves die back
- Leaves turn light green to yellow
- Stalks become short and slender
- Vegetative growth reduced
- Tips and margins of older leaves become necrotic

Phosphorus deficiency

- ➢ Leaf blades turn dark green to blue green
- ➢ Tips and margins show red or purple colouration (due to direct exposure to sun light)
- ➢ Slender leaves; older leaves turn yellow and die back from tips and along margins
- ➢ Stalks short and slender
- ➢ Internodal length reduced towards the tip
- ➢ Plant vigour and tillering reduced

Potassium deficiency

- ❖ Older leaves entirely brown or 'fired' with necrotic lesions between veins along margins and tips
- ❖ Red discolouration on the upper surface of leaf midrib
- ❖ Leaf spindle distort producing 'bunched top' or 'fan' appearance
- ❖ Stalks slender

Boron deficiency

- ➢ Younger leaves distorted and necrotic with translucent lesions between veins
- ➢ Leaf tips burned, young plants bunched with many tillers
- ➢ Leaves brittle, chlorotic and later necrotic

Molybdenum deficiency

- ❖ Older leaves dry and die back
- ❖ Longitudinal chlorotic streaks
- ❖ Vegetative growth reduced
- ❖ Stalks short and slender

Magnesium deficiency

- ➢ Tips and margins of older leaves first become mottled or necrotic

- ➢ Red necrosis and hence rusty appearance
- ➢ Internal browning of rind

Calcium deficiency

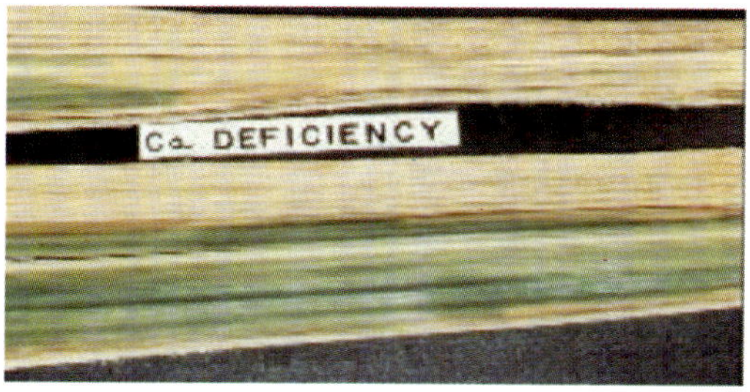

- ❖ Older leaves rusty, hook or turn downward
- ❖ Young leaves distorted, hook or turn downward and necrotic
- ❖ Apical meristems die
- ❖ Chlorotic lesions with necrotic spots on young leaves, later turn reddish brown
- ❖ Rind soft and stalks slender

Sulphur deficiency

- ➢ Youngest leaves first affected, become chlorotic
- ➢ Stalks slender

Iron deficiency

- ❖ Young leaves distorted and chlorotic
- ❖ Interveinal chlorosis from tip to base of leaves
- ❖ Entire plant chlorotic

Zinc deficiency

- ➢ Leaves small and non symmetrical
- ➢ Veinal chlorosis first at base of young leaves
- ➢ Necrosis under severe deficiency
- ➢ Reduced tillering and ratooning ability
- ➢ Younger leaves distorted and necrotic with translucent lesions between veins

- ➢ Leaf tips burned; young plants bunched with many tillers
- ➢ Leaves brittle, chlorotic and later necrotic

Copper deficiency

- ❖ Green splotches on leaf blades
- ❖ Leaves papery thin and rolled
- ❖ Stalks non-turgid and rubbery

Manganese deficiency

- ❖ Interveinal chlorosis occurs from tip to middle leaf
- ❖ Chlorotic stripes bleach and fray in the wind

2. SUGAR BEET

Nitrogen deficiency

- ➢ Leaves pale green, turning dark reddish purple; later growth thin and erect

Phosphorus deficiency

- ❖ Leaves dull purple, small with short petioles

Potassium deficiency

- ➢ Leaves bluish green around veins; surface crinkled and margins curled downward; older leaves become yellowish to reddish brown between veins and at margins; stems of older leaves show brown spots and stripes; roots are dark and tend to rot and poorly developed

Calcium deficiency

- ❖ Leaves pale green around margins, curled toward upper surface, necrotic and ragged.

❖ Roots are forked and turned

Magnesium deficiency

➢ Older leaves show chlorosis and reddish tinting between veins

Manganese deficiency

❖ Leaves chlorotic between veins, with erect growth, margins are curled towards upper surfaces;

❖ Red and purple tinting of leaves

Molybdenum deficiency

➢ Red veins more conspicuous against chlorotic background

F. FRUITS

1. CITRUS

Nitrogen deficiency

❖ When the deficiency is chronic, first mature leaf turns yellow and abscises, leaving only young, pale green terminal leaves on the tree

❖ Trees may have thin foliage, twigs die back, limited fruit production with inferior quality

❖ Under mild deficiency, premature shedding of older leaves is reduced and only a small decline in fruit yield and quality

Phosphorus deficiency

➢ Reduced yield, premature defoliation.

➢ Smaller size of new leaves that develop a point

- ➢ Fruit quality affected; coarse thick peels, pale colour, spongy fruit and excessive acidity of juice

Potassium deficiency

- ❖ Small crinkled leaves that may develop yellow to bronze chlorosis in the outer half of leaf later chlorotic leaf tips turn brown giving an appearance of scorch or "tip burn"
- ❖ Reduced fruit size, thinner peels, lower juice acidity with improved fruit color and higher juice content

Calcium deficiency

- ➢ Stunted growth of tree, poor root development, dieback of twigs
- ➢ Yellowing of leaf margin and interveinal areas

Magnesium deficiency

- ❖ Mature leaves show symptoms of bronzing and yellowish blotches between the main veins and along the midribs
- ❖ Margins of leaves fade inward leaving an inverted green " V " at the leaf base

Sulphur deficiency

- ➢ Deficiency symptoms similar to those of N deficiency

Iron deficiency

- ❖ Only young leaves affected
- ❖ Yellow leaves with green midrib and lateral veins
- ❖ In moderate to severe cases, leaves may be solid yellow but persistent
- ❖ In chronic cases, growth is restricted and dieback occurs

Remarks

- ➢ Fe deficiency mainly occurs on calcareous soils

Zinc deficiency

- ❖ Leaf size reduced and a distinctive yellow mottle appears on leaves
- ❖ When deficiency prolongs, many shoots affected, leaves become very narrow and short and almost completely yellowed
- ❖ Continuous deficiency leads to defoliation and die back of twigs, fruits small and misshapen and bleached to whitish color
- ❖ When supply of zinc is gradually restored, green bands along the leaf veins appear

Remarks

- ❖ Zn and Mn or Zn and Fe deficiencies often coexist on the same shoots and the resulting chlorosis may be confusing

Manganese deficiency

- ➢ Deficiency occurs in young emerging leaves
- ➢ Leaves show a light green mottle on a dark back ground, the pattern resembling a series of inverted "horse shoes" around the periphery of the leaf. At times it may be a complete marbling of light green or dark green. If Zn or Mg deficiency or anion toxicity coexist, yellowish, brownish coloured pattern may occur

Copper deficiency

- ❖ Leaves abnormally large, dark green with no chlorosis
- ❖ Twigs dieback, and multiple buds form near the juncture with live wood
- ❖ Most reliable symptoms for identification are gum pockets under the green bark of young wood and brownish excrescences on fruit twigs and leaves
- ❖ New shoots narrow, leaves elongated similar to peach leaves
- ❖ Under more severe deficiency, fruit develops brown lesions, the rigid peel may split and fruit drops

Boron deficiency

- ➢ Deficiency symptoms are vague
- ➢ Short internodes drab colour, yellow veins similar to girdling, thickened leaves, corky venation and buckling of leaf blade along midrib

- Symptoms identifiable only under severe deficiency
- Continued deficiency of B leads to a collapse of phloem which produces symptoms similar to mechanical girdling or girdling produced by diseases

Molybdenum deficiency

- Matured leaf develops "yellow spot"; spot occurs randomly on the leaf and gradually become gums impregnated, turn reddish, brown and bulge in the underside of the leaf
- Premature fall of leaves presenting a thinly foliated tree

2. GRAPES

Nitrogen deficiency

- Leaf blades turn pale green and then yellow
- Young shoots, petioles and cluster stems become pink or red
- Growth of the shoots slowed down
- Fruits mature early and do not reach normal size

Remarks

- Water shortage and root or trunk injury may also lead to stunted growth and pale green leaves which may be confused with nitrogen deficiency

Phosphorus deficiency

- Shoot and root growth reduced
- Leaves small and dark green, under severe deficiency leaves turn reddish

Potassium deficiency

- During the early part of the season, leaves lighter in color with a few necrotic spots along the margin of young blades
- Leaf margins dry and role upwards or downwards and blades become distorted and ruffled

- ❖ Shiny round or elliptical spots appear on the upper side of mature leaves; when the weather is sunny and dry; such spots unite and become larger light yellow shiny areas that turn brown within a few days
- ❖ Surfaces of older leaves at the base of shoots become lilac brown or dark brown (black leaves). The discoloration starts in the internal tissues and spreads over the entire leaf blades; under severe deficiency, the leaf margins role upwards

Calcium deficiency

- ➢ A narrow, necrotic border first appears at the leaf margins and then moves towards the petioles, and in extreme cases covers the entire leaf blades
- ➢ The growing clusters dry up starting from the tip

Magnesium deficiency

- ❖ If deficiency is before bloom, interveinal and marginal leaf necrosis prominent
- ❖ During summer, interveinal yellowing or reddening and development of brown green patches are the major symptoms
- ❖ The discoloration increases from margin towards petiole in a wedge shape

Remarks

- ➢ Mg deficiency occurs on light, acid soils. High application of K or ammonium also induces magnesium deficiency. When both Ca and Mg are deficient, there will be acidity damage and stem necrosis.

Iron deficiency

- ❖ When iron chlorosis appears, severe fading of leaves at the margins and from there yellowing enters the interveinal areas
- ❖ Edges of yellowed leaves dry and dead tissues roll upwards; finally the leaves dry and fall off
- ❖ Lateral shoots develop with short, thin, pink coloured internodes
- ❖ A small bleached, narrow leaflets develop and then fold along the central vein

- ❖ When chlorosis is severe, leaflets dry up and the shoot desiccates starting from the tip
- ❖ The reddish coloured shoots have thickened nodes that grow yellowed tendrils
- ❖ The flower clusters also become yellow
- ❖ After bloom, poor fruit set and in extreme cases all berries are lost due to shatter (dropping of unpollinated flowers and fruits)

Zinc deficiency

- ➢ Small leaf blades with wide opened petiole have sinuses and sharp teeth; they are asymmetrical, one half of the leaf is always larger than the other
- ➢ Small berries and loose clusters

Remarks

- ❖ When P content is high, Zn will combine with phosphate to form water - insoluble Zinc phosphate so that roots cannot absorb it

Manganese deficiency

- ➢ Leaves at the base of the shoot start to pale and then small polygonal yellow spots appear as the interveinal brown colour; this symptom may be confused with the infection of Plasmopara viticola (brown rot)

Remarks

- ❖ Mn deficiency occurs mainly in alkaline sandy soils containing high amounts of humus

Boron deficiency

- ➢ Roots short, thick, swollen into knots and break open longitudinally

3. BANANA

Nitrogen deficiency

- ❖ Stunted growth, reduced leaf size, paling of leaf colour (later with marginal leaf chlorosis)

- ❖ Short thin petioles
- ❖ Reduced suckering

Phosphorus deficiency

- ➢ Stunting due to slower rate of leaf production
- ➢ Leaves some what choked, initially green, later on chlorotic along margins leading to premature death

Potassium deficiency

- ❖ Small reduction of growth
- ❖ Leaf size reduced, chlorosis spreading from tip towards the base
- ❖ Leaf colour turns orange yellow accompanied by marginal and distal scorching which proceeds rapidly and irregularly until the whole leaf is dry
- ❖ The dead leaf remains attached or breaks near the bottom of the lamina, rather than at the base of the petiole and this is a characteristic symptom
- ❖ Under mild deficiency, premature yellowing of leaves with thin, oblique misshapen branches

Calcium deficiency

- ➢ Older leaves exhibit marginal flecking and chlorosis
- ➢ Lateral veins of the younger leaves thickened to be followed by marginal necrosis

Magnesium deficiency

- ❖ Older leaves pale in colour with necrotic flecking; flecks later acquire yellow halos
- ❖ Sometimes green strips along the margins and midribs

❖ Under severe cases, stunting, chocking and breaking away of leaf sheaths

Sulphur deficiency

➢ First, younger leaves show general chlorosis

➢ When deficiency is acute, new leaves remain chlorotic and yellowing progresses to other leaves until the midrib alone remains green

Iron deficiency

❖ Young leaves exhibit interveinal chlorosis

Manganese deficiency

➢ Marginal interveinal chlorosis in the younger leaves leading to coalescent necrotic spotting and finally a necrotic brown leaf margin

➢ Sometimes plants show black pimply spots

Zinc deficiency

❖ Severe deficiency produces narrow, tapered and chlorotic leaves

❖ Mild deficiency has the effect of nitrogen shortage

❖ Young leaves bunched or chocked

❖ Bunches small and misshapen with small, twisted, chlorotic bottle necked fruits

Copper deficiency

➢ Yellowing and drooping of leaves

4. APPLE

Nitrogen deficiency

❖ Small; pronounced yellow and red tints.

Phosphorus deficiency

➢ Small; dull purple and bronze tints

Magnesium deficiency

❖ Severe defoliation of terminal shoots progressing from base to tip.

- ❖ Intervenal chlorosis and necrosis.

Potassium deficiency

- ➢ Bluish green, with slight marginal and interveinal chlorosis, followed by marginal scorching, either brown or grayish brown color.

Iron deficiency

- ❖ Tip leaves chlorotic
- ❖ Small veins show as fine network in early stages
- ❖ Margins develop brown patches.

Manganese deficiency

- ➢ Leaves show severe chlorosis over most of tree
- ➢ Young leaves of terminal shoots not as severely affected as older leaves.

- ❖ Leaves interveinal chlorosis progressing from margins towards midrib.

Zinc deficiency

- ➢ Buds along shoots fail to develop
- ➢ Leaves small and narrow ("Little Leaf" condition) and tend to form rosettes at tips of shoots.

5. BERRIES

a. GOOSEBERRY

Phosphorus deficiency

- ❖ Leaves show dull purple tints becoming red later.

Magnesium deficiency

- ➢ Leaves possess broad red marginal bands which fade to cream color.
- ➢ Broad red marginal bands, later fading to cream.

Potassium deficiency

- ❖ Bluish green color, margins generally curled backward

- ❖ Develop purple tints followed by grayish brown marginal scorch
- ❖ Purpling disappears as scorching develops.

b. RASPBERRY

Potassium deficiency

- ➢ Marginal leaf scorch and interveinal browning
- ➢ Leaves tend to curl backward towards under surfaces, but scorched margins turn forward.

Manganese deficiency

- ❖ Leaves interveinal chlorosis beginning near margins and progressing towards midribs as well defined pattern

❖ Terminal leaves green or less chlorotic than older leaves, in contrast to iron deficiency.

Iron deficiency

➢ Terminal leaves severe chlorosis followed by marginal and interveinal necrosis.

c. STRAW BERRY

Iron deficiency

❖ Leaves severe chlorosis; green network of fine veins distinct in early stages.

Manganese deficiency

➢ Somewhat faint interveinal chlorosis beginning at margins and progressing towards midrib; where severe, chlorotic areas yellowish appearance.

6. MANGO

Potassium deficiency

❖ Mature leaves show drying of tips and margins

Iron deficiency

- Young leaves chlorotic; during early stages, leaf veins will remain green and later on entire leaf will become chlorotic

Boron deficiency

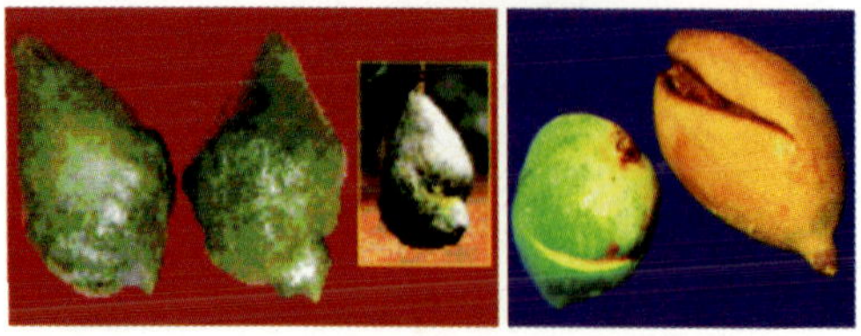

- Fruits become irregularly shaped
- Cracks will also be seen on the fruit

Zinc deficiency

- Leaves small and narrow
- Rosetting of leaves at tips of shoots

Copper deficiency

- Tips of branches dry up
- Fruits display dots and eruptions on their surface

7. POMEGRANATE

Nitrogen deficiency

- Leaves become uniformly chlorotic yellow with light green midrib and veins
- In extreme cases, necrotic spots develop on the chlorotic margins of the leaves; few branches develop
- Leaf size reduced and defoliation occurs

Phosphorus deficiency

- Overall suppression of growth; Shoots and stems become thin, number of branches and leaves are reduced, leaves of middle and apical portions on tertiary and side shoots become chlorotic yellow
- Basal older leaves remain dull green and lose luster
- Chlorosis starts at margins and tips of the leaves and the margins of young leaves roll

Potassium deficiency

- Reduction in size and number of leaves
- Leaves at the middle develop chlorotic symptoms on the margins with few patches on the lamina having rusty brown or blackish grey spots
- Backward curling of leaves from tips to base

Magnesium deficiency

- Overall suppression of growth
- Interveinal chlorosis of basal leaves with green midribs and veins, necrosis of the chlorotic area
- Inverted 'V' shape of basal chlorotic scorched leaf

Sulphur deficiency

- Suppression of growth

- Plant stunted with few branches and leaves
- Heavy defoliation occurs
- Younger leaves become chlorotic yellow
- Necrotic spots develop on the margins of the chlorotic leaves

Calcium deficiency

- Chlorotic symptoms and termination of apical growth of shoots, sparse foliage and hooking of leaf tips
- Terminal leaves become chlorotic yellow and develop whitish patches

G. VEGETABLES

1. BRINJAL

Nitrogen deficiency

- Plants remain stunted and unbranched and turn pale green
- Leaves are markedly smaller in size and stiff in texture.
- Older leaves gradually start getting bleached from margin inwards to pale in colour, and finally shed prematurely

Phosphorus deficiency

- Plant growth is retarded
- Leaves are smaller in size and as deficiency progresses, leaves turn dirty green
- Fruits small, pale in colour and mature early

Potassium deficiency

- Growth of the plant suppressed
- Marked reduction in the number and size of leaves
- Brown patches appear on the mature leaves
- Small necrotic lesions develop along the veins, followed by defoliation

Magnesium deficiency

- Symptoms of Mg deficiency will be marked
- Stunting of growth
- Lower leaves develop a characteristic interveinal chlorosis
- Chlorosis from margins inward in a typical inverted "V" shape
- Leaves ultimately turn pale yellow

2. CHILLIES

Nitrogen deficiency

- Deficiency of nitrogen results in suppression of growth
- Branches are short, thin and few in number with a few short and distorted leaves
- Colour of the leaves gradually changes from light green to yellowish green which drop off prematurely

Phosphorus deficiency

- Deficient plants are usually stunted in growth
- Leaves are small, narrow and inwardly curved
- Older leaves are yellow with pink margins

Potassium deficiency

- ➢ The growth is suppressed
- ➢ Leaf number is greatly reduced and colour becomes yellowish
- ➢ Crinkling of leaves with necrotic spots and marginal scorching

- ➢ Plants stunted
- ➢ Leaves small and become twisted and discolored

3. CARROT

Nitrogen deficiency

- ❖ Leaves coloured light green, followed by yellowing; petioles weak

Phosphorus deficiency

- ➢ Leaves are dull green with purple colouration, older leaflets die, petioles are upright

Potassium deficiency

- ❖ Leaves are slightly chlorotic followed by browning; roots are spindly; growth is short

Calcium deficiency

- ➢ Leaves show some chlorosis, necrosis, or scorching and finally die, foliage is sparse

Magnesium deficiency

- ❖ Leaves are light coloured; tips or lobes show light yellow or brown spots, older leaves are severely chlorotic

Molybdenum deficiency

- ➢ Leaves show marginal yellowing, followed by red extending inward
- ➢ Roots have wide deep splits

4. POTATO

Nitrogen deficiency

- ❖ Leaves light green; older or lower leaves turn upright; tubers small

Phosphorus deficiency

- ➢ Leaves show forward roll and marginal scorch; the older ones drop; growth is upright and spindly; tubers may have internal lesions

Potassium deficiency

- ❖ Leaves are first bluish green, older leaves become yellowish followed by necrosis and browning starting from tips and margins, leaflets are cupped and crowded together; stalks are slender with short internodes and may collapse prematurely; tuber flesh is bluish

Calcium deficiency

- ➢ Young leaves are small and pale green, rolled towards upper surfaces, with marginal necrosis; bud dies; tubers are dwarfed and useless, with dead spots in pith region

5. BHENDI

Nitrogen deficiency

- Stunted growth, thin shoots small and sickly leaves;
- Abscission of leaves at advanced stage of deficiency, the plants ultimately turning yellow with hard and brittle foliage, premature bolting with small and rough pods

Phosphorus deficiency

- Stunted growth with dark foliage

Potassium deficiency

- Suppression of leaf formation turning brown yellow, followed by brown colouration and scorching and abscission of scorched leaves

Magnesium deficiency

- Interveinal yellow spots appear on leaves; these increase in number until only the veins remains green

6. CUCURBITS

Nitrogen deficiency

- Vegetative growth reduced, yellowing of leaves first in mature leaves and then in younger leaves
- Size of the foliage reduced
- Fruit set reduced and fruits formed will be smaller
- High incidence of cull fruits

Phosphorous deficiency

- Stunted appearance and slower rate of growth
- Under cool growing seasons, leaves may be purple

Potassium deficiency

- Rate of growth very slow
- Tips of the older leaves chlorotic and necrotic
- Fruit development irregular, small at the stem end and normal at the blossom end

Calcium deficiency

- New root growth impaired
- Young leaves reduced in size, distorted and spotted and necrotic at the leaf tips
- Under severe deficiency, terminal buds die
- Blossom end of the fruit wall thinner and water soaked, with brown necrotic spots

Magnesium deficiency

- ❖ First chlorosis occurs on older leaves at the leaf tips and progress towards the interveinal area
- ❖ Under severe deficiency, fruit development and yield are reduced

Remarks

- ➢ Deficiency occurs most often on coarse textured, acidic soils in humid regions; high concentrations of potassium, calcium and ammonium may induce Mg deficiency

Sulphur deficiency

- ❖ Symptoms first appear on the young leaves and then other leaves
- ❖ Plants stunted, veins become slightly lighter in colour

Iron deficiency

- ➢ Younger leaves become chlorotic interveinally. If deficiency persists, entire plant is affected with development of necrotic spots

Zinc deficiency

- ❖ Reduced leaf size, interveinal chlorosis that later becomes necrotic, and shortened internodes
- ❖ Plant growth reduced

Remarks

- ➢ Zinc deficiencies are more severe under cool, wet, under cloudy situations; excess of P, Fe, and Cu may induce Zn deficiency

Manganese deficiency

- ❖ Interveinal area of younger leaves become chlorotic while veins remain green
- ❖ Afterwards older leaves affected

Remarks

- ➢ Manganese deficiency may be induced by over liming

Copper deficiency

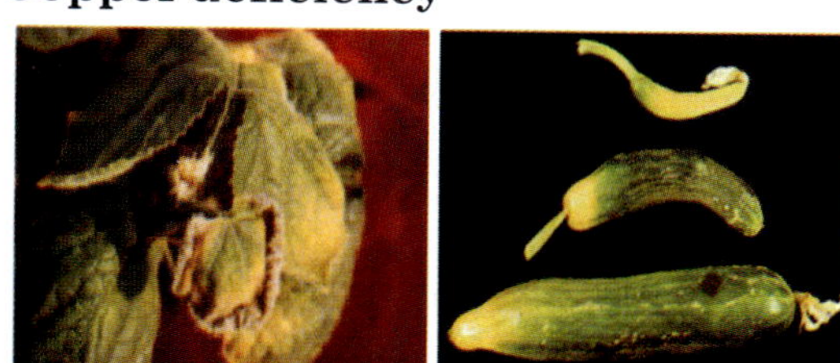

- ❖ Growth reduced, leaves wrinkled, internodes shortened
- ❖ Leaves chlorotic and necrotic

- ❖ At later stages, younger leaves distorted and apical may die back

Boron deficiency

- ➢ Symptoms first appear on young foliage; shortened internodes and distorted leaves that become chlorotic and then necrotic, with the death of terminal buds
- ➢ Fruit affected by cracks, necrotic spots and internal breakdown

Molybdenum deficiency

- ❖ Symptoms similar to those of nitrogen deficiency; older leaves pale green with some interveinal mottling and later become necrotic
- ❖ Fruit set reduced and younger leaves become cupped and distorted

Remarks

- ➢ Deficiencies of Mo are observed on highly acidic soils

7. TOMATO

Nitrogen deficiency

- ❖ Growth dwarfed, thin and upright habit
- ❖ Stem and petioles rigid
- ❖ Leaves pale green, occasional purplish tint, older leaves yellowing.

Phosphorus deficiency

- ➢ Growth dwarfed and thin;
- ➢ Leaflets droop, curl backward and develop strong, dull purple tints.

Potassium deficiency

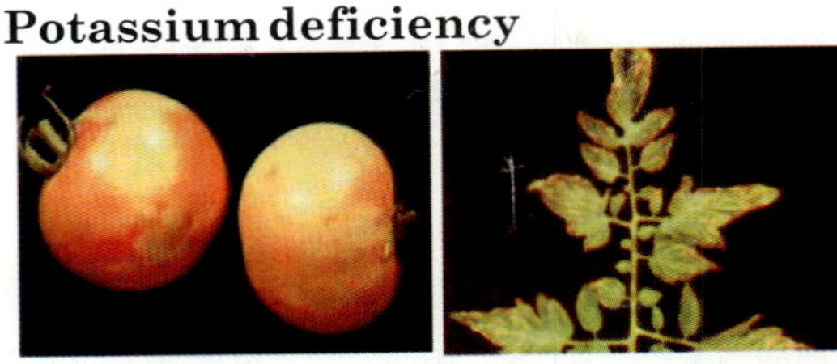

- ❖ Leaflets slight marginal and interveinal chlorosis followed by brown marginal scorching;
- ❖ Scorched margins curled forward.
- ❖ "Blotchy Ripening"; green and yellow areas merging into red colouration on fruit skin.

Magnesium deficiency

- ➢ Central interveinal chlorosis and green marginal bands.

Calcium deficiency

- ❖ Drying off of terminal leaflets and flowers; leaves purplish brown tinting.
- ❖ Die back of tissues and "Blossom End Wilt" of distal fruitlets

Iron deficiency

- ➢ Tip leaves, especially basal areas of leaflets, intense chlorotic mottling
- ➢ Stem near tip also yellow.

Boron deficiency

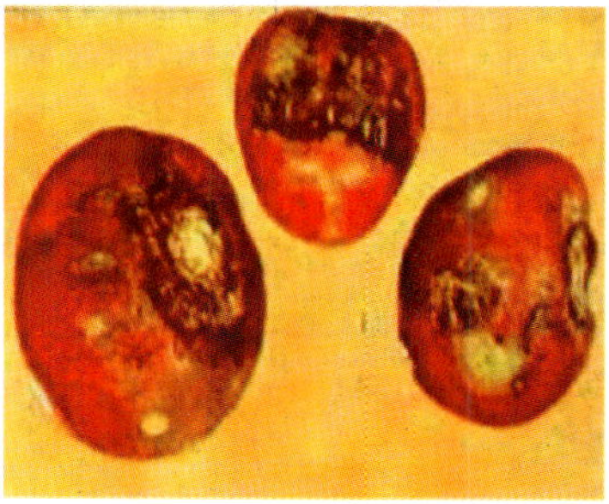

- ❖ Stems stiff; terminal buds die and growths die back.
- ❖ Lateral shoots developed, giving plant flat top
- ❖ Leaves highly tinted purple, brown and yellow
- ❖ Fruits pitted and corky areas in skin; ripening uneven.

Molybdenum deficiency

- ➢ Leaflets somewhat chlorotic, strongly incurled and die back from tips.
- ➢ Incurled margins, interveinal chlorotic mottling and death of tips.

8. CAULIFLOWER

Nitrogen deficiency

- ➢ Growth dwarfed, erect and stem thin; leaves pale green
- ➢ Older leaves bright orange, red and purple tints.

Calcium deficiency

- ❖ Hooking of young leaves and severe collapse of tissues.

Magnesium deficiency

- ➢ Older leaves chlorotic marbling followed by strong orange, red and purple tinting.

Potassium deficiency

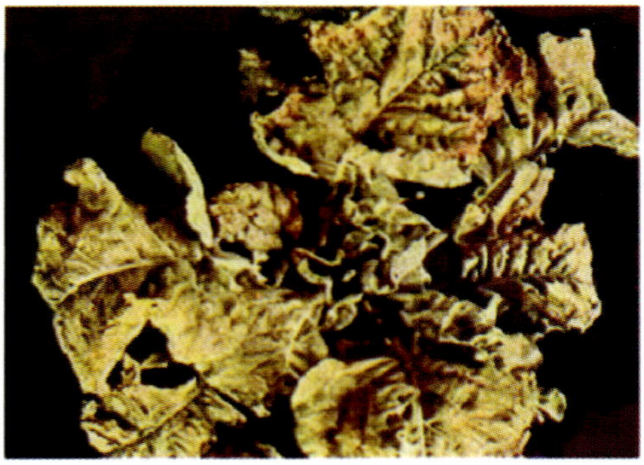

- ❖ Leaves bluish green and slight interveinal chlorosis near margins;
- ❖ Backward curling of laminae; scorching and forward curling of margins.

Boron deficiency

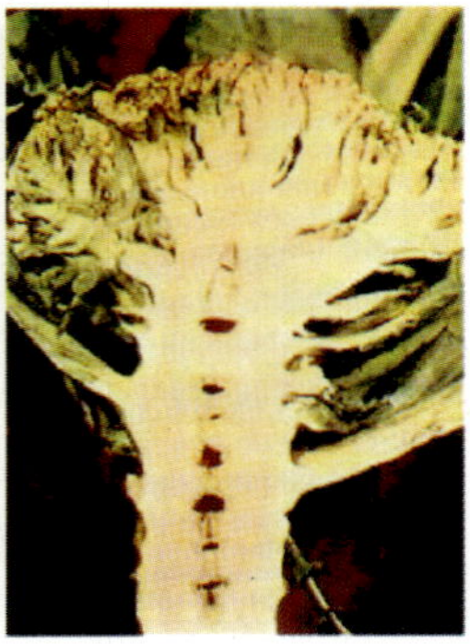

- ➢ Browning of curd
- ➢ Lesions in pith
- ➢ Corky condition of epidermis.

Molybdenum deficiency

- ❖ Leaves slightly cupped and show chlorotic mottling
- ❖ tips develop dead patches; cotyledons remain dark green.
- ❖ Laminae of leaves restricted and may fail to develop, leaving only bare midribs.

9. CABBAGE

Phosphorus deficiency

- ➢ Older leaves dull purple tints.

Calcium deficiency

- ❖ Older leaves, marginal and interv-einal collapse
- ❖ Young leaves, cupping and distort-ion and scorching of margins.

Magnesium deficiency

- ❖ Older leaves chlorotic marbling.

Potassium deficiency

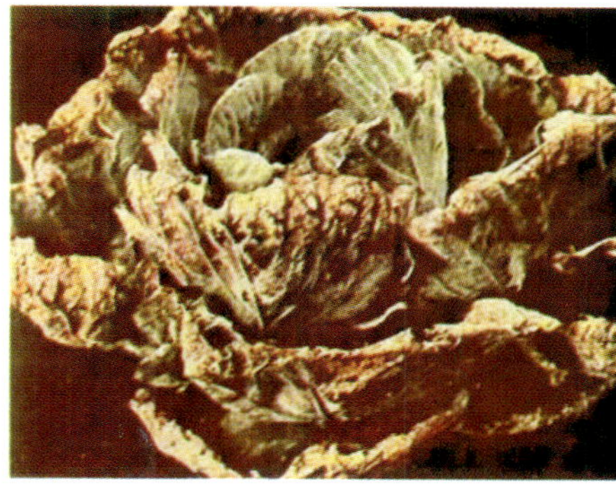

- ➢ Older leaves severe marginal scorch and forward curling of leaf margins.

Iron deficiency

- ❖ Severe chlorosis of leaves beginning as a chlorotic mottling.

Manganese deficiency

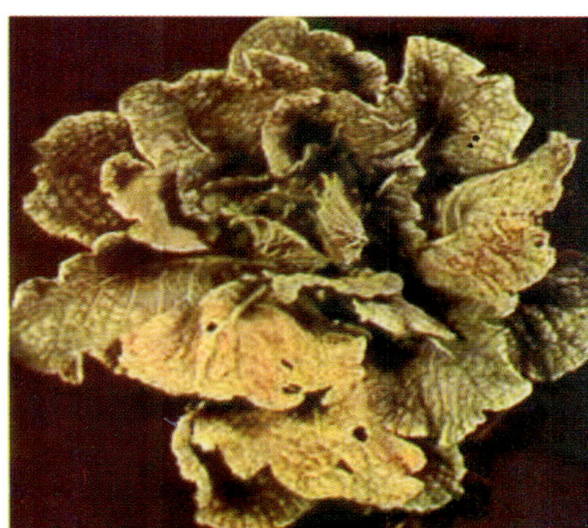

- Leaves show severe interveinal chlorotic marbling and necrosis.

Boron deficiency

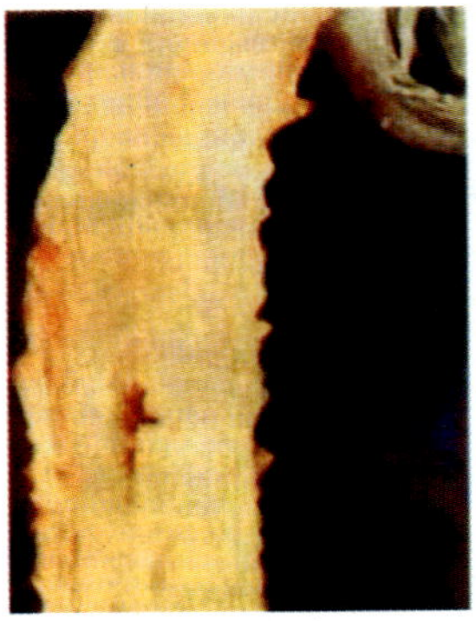

- Longitudinal section showing necrosis and splitting of pith.

Molybdenum deficiency

- Leaves cupped and show chlorotic mottling, especially around margins
- Tips and margins develop dead patches

10. ONION

Potassium deficiency

- Die back of leaves from tips and drying off of older leaves.

Nitrogen and phosphorus deficiencies

- Left: nitrogen deficient, growth dwarfed and thin, leaves pale green, tip die-back.
- Right: phosphorus deficient, growth dwarfed and thin, leaves dull green, tip die-back.

Calcium deficiency

- Growth dwarfed; leaves chlorosis developing from tips followed by die-back.

11. RADISH

Phosphorus deficiency

- Growth dwarfed; leaves lustreless dark green, laminae and petioles dull purple tints;
- Older leaves marginal scorch and wither early.

Magnesium deficiency

- Older leaves marked interveinal chlorosis.

Potassium deficiency

- Leaves backward curling and scorching of margins.

H. FLOWERS

1. ROSE

Nitrogen deficiency

- Old leaves become smaller
- Reddish spots present on leaves
- Stems thin and few
- Poor development of buds
- Flowers small and light coloured

Phosphorus deficiency

- Leaves dark green.
- Leaves may drop before turning yellow.
- Stems weak
- Poor root development

Potassium deficiency

- Marginal browning and scorching of lower leaves
- Stems weak and growth stunted
- Buds fail to open and flower colour is pale
- Short flowering starts with smaller flowers

Magnesium deficiency

- Interveinal chlorosis of older leaves
- Reduced number of stems
- Flowers smaller and poor in colour
- Reduced root growth and flower yield

Calcium deficiency

- Hard and stubby growth of plant

- ❖ Root system short and thickened
- ❖ Terminal buds die
- ❖ Defoliation of leaves
- ❖ Margins of leaflets die and rest of the leaf turns yellowish green
- ❖ Dark brown patches appear between midrib and margin
- ❖ Leaves drop off before they are dead
- ❖ Flowers deformed, reduced in number and have shown spots near margin of petals

Iron deficiency

- ➢ Interveinal chlorosis of terminal leaves.
- ➢ Younger leaves more chlorotic and transparent

Boron deficiency

- ❖ Terminal buds die and hence excessive branching
- ❖ Leaves crowded together at the tips of the stems giving a tufted appearance
- ❖ Flowers bull headed or malformed
- ❖ Acute deficiency causes death of stem tips

Copper deficiency

- ➢ Plants dwarf and bushy
- ➢ Leaves narrower and twisted.
- ➢ Leaf tips become white

Manganese deficiency

- ❖ Interveinal chlorosis
- ❖ Deformities in tissue formation

Molybdenum deficiency

- ➢ Shortened plant growth
- ➢ Leaves pale and wither
- ➢ Symptoms resemble those of nitrogen deficiency

Sulphur deficiency

- ❖ General yellowing of foliage

2. JASMINE

Iron deficiency

- ➢ Leaves exhibit interveinal chlorosis

Magnesium deficiency

- ❖ First chlorosis occurs on older leaves at the leaf tips and progress towards the interveinal area
- ❖ Under severe deficiency, fruit development and yield are reduced

Remarks

➢ Deficiency occurs most often on coarse textured, acidic soils in humid regions; high concentrations of potassium, calcium and ammonium may induce Mg deficiency

Sulphur deficiency

- ❖ Symptoms first appear on the young leaves and then other leaves
- ❖ Plants stunted, veins become slightly lighter in colour

Iron deficiency

➢ Younger leaves become chlorotic interveinally. If deficiency persists, entire plant is affected with development of necrotic spots

Zinc deficiency

- ❖ Reduced leaf size, interveinal chlorosis that later becomes necrotic, and shortened internodes
- ❖ Plant growth reduced

Remarks

➢ Zinc deficiencies are more severe under cool, wet, under cloudy situations; excess of P, Fe, and Cu may induce Zn deficiency

Manganese deficiency

- ❖ Interveinal area of younger leaves become chlorotic while veins remain green
- ❖ Afterwards older leaves affected

Remarks

➢ Manganese deficiency may be induced by over liming

Copper deficiency

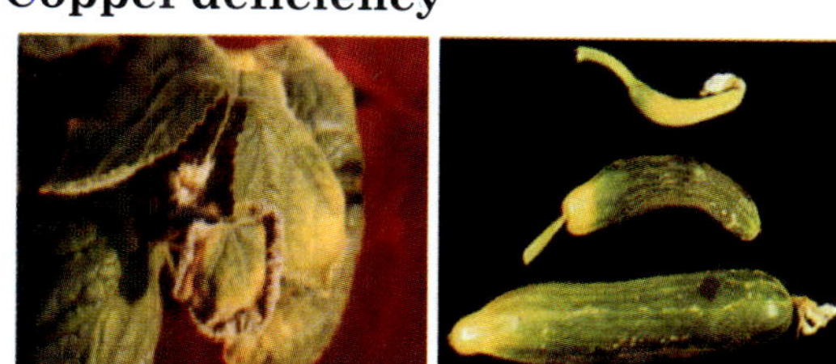

- ❖ Growth reduced, leaves wrinkled, internodes shortened
- ❖ Leaves chlorotic and necrotic

- At later stages, younger leaves distorted and apical may die back

Boron deficiency

- Symptoms first appear on young foliage; shortened internodes and distorted leaves that become chlorotic and then necrotic, with the death of terminal buds
- Fruit affected by cracks, necrotic spots and internal breakdown

Molybdenum deficiency

- Symptoms similar to those of nitrogen deficiency; older leaves pale green with some interveinal mottling and later become necrotic
- Fruit set reduced and younger leaves become cupped and distorted

Remarks

- Deficiencies of Mo are observed on highly acidic soils

7. TOMATO

Nitrogen deficiency

- Growth dwarfed, thin and upright habit
- Stem and petioles rigid
- Leaves pale green, occasional purplish tint, older leaves yellowing.

Phosphorus deficiency

- Growth dwarfed and thin;
- Leaflets droop, curl backward and develop strong, dull purple tints.

Potassium deficiency

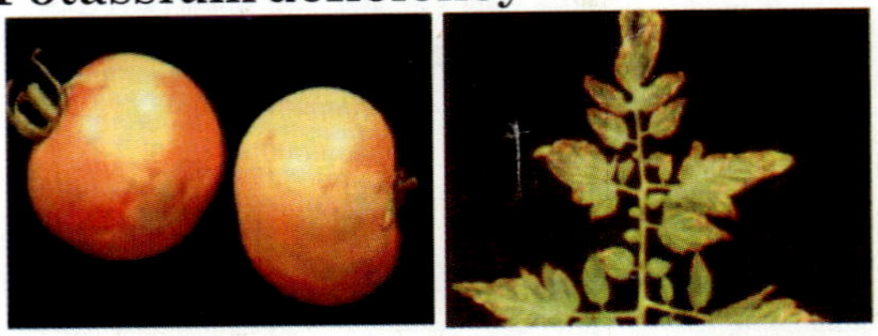

- Leaflets slight marginal and interveinal chlorosis followed by brown marginal scorching;
- Scorched margins curled forward.
- "Blotchy Ripening"; green and yellow areas merging into red colouration on fruit skin.

Magnesium deficiency

- Central interveinal chlorosis and green marginal bands.

Calcium deficiency

- ❖ Drying off of terminal leaflets and flowers; leaves purplish brown tinting.
- ❖ Die back of tissues and "Blossom End Wilt" of distal fruitlets

Iron deficiency

- ➢ Tip leaves, especially basal areas of leaflets, intense chlorotic mottling
- ➢ Stem near tip also yellow.

Boron deficiency

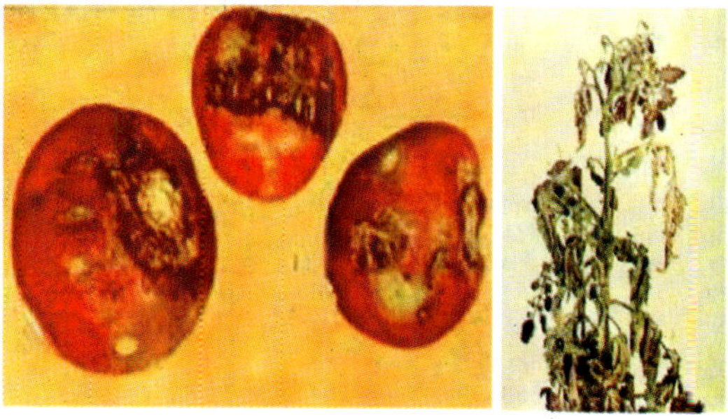

- ❖ Stems stiff; terminal buds die and growths die back.
- ❖ Lateral shoots developed, giving plant flat top
- ❖ Leaves highly tinted purple, brown and yellow
- ❖ Fruits pitted and corky areas in skin; ripening uneven.

Molybdenum deficiency

- ➢ Leaflets somewhat chlorotic, strongly incurled and die back from tips.
- ➢ Incurled margins, interveinal chlorotic mottling and death of tips.

8. CAULIFLOWER

Nitrogen deficiency

- ➢ Growth dwarfed, erect and stem thin; leaves pale green
- ➢ Older leaves bright orange, red and purple tints.

Calcium deficiency

- ❖ Hooking of young leaves and severe collapse of tissues.

Magnesium deficiency

- ➢ Older leaves chlorotic marbling followed by strong orange, red and purple tinting.

Potassium deficiency

- ❖ Leaves bluish green and slight interveinal chlorosis near margins;
- ❖ Backward curling of laminae; scorching and forward curling of margins.

Boron deficiency

- ➢ Browning of curd
- ➢ Lesions in pith
- ➢ Corky condition of epidermis.

Molybdenum deficiency

- ❖ Leaves slightly cupped and show chlorotic mottling
- ❖ tips develop dead patches; cotyledons remain dark green.
- ❖ Laminae of leaves restricted and may fail to develop, leaving only bare midribs.

9. CABBAGE

Phosphorus deficiency

- ➢ Older leaves dull purple tints.

Calcium deficiency

- Older leaves, marginal and interveinal collapse
- Young leaves, cupping and distortion and scorching of margins.

Magnesium deficiency

- Older leaves chlorotic marbling.

Potassium deficiency

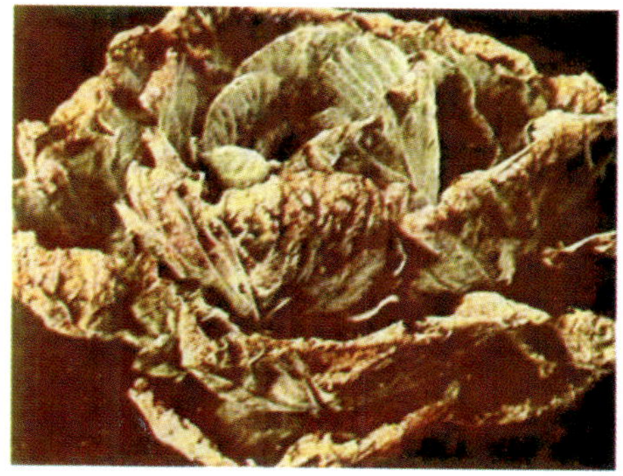

- Older leaves severe marginal scorch and forward curling of leaf margins.

Iron deficiency

- Severe chlorosis of leaves beginning as a chlorotic mottling.

Manganese deficiency

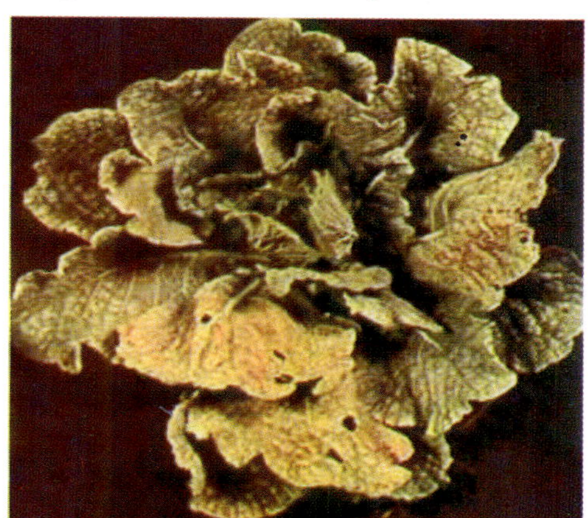

➢ Leaves show severe interveinal chlorotic marbling and necrosis.

Boron deficiency

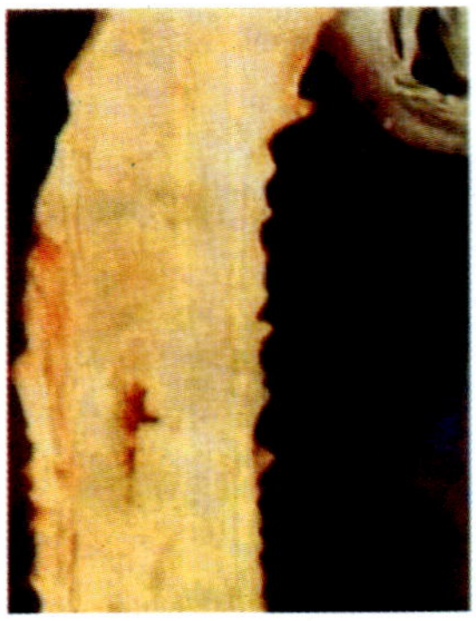

❖ Longitudinal section showing necrosis and splitting of pith.

Molybdenum deficiency

➢ Leaves cupped and show chlorotic mottling, especially around margins

➢ Tips and margins develop dead patches

10. ONION

Potassium deficiency

❖ Die back of leaves from tips and drying off of older leaves.

Nitrogen and phosphorus deficiencies

➢ Left: nitrogen deficient, growth dwarfed and thin, leaves pale green, tip die-back.

➢ Right: phosphorus deficient, growth dwarfed and thin, leaves dull green, tip die-back.

Calcium deficiency

❖ Growth dwarfed; leaves chlorosis developing from tips followed by die-back.

11. RADISH

Phosphorus deficiency

➢ Growth dwarfed; leaves lustreless dark green, laminae and petioles dull purple tints;

➢ Older leaves marginal scorch and wither early.

Magnesium deficiency

❖ Older leaves marked interveinal chlorosis.

Potassium deficiency

➢ Leaves backward curling and scorching of margins.

H. FLOWERS

1. ROSE

Nitrogen deficiency

❖ Old leaves become smaller

❖ Reddish spots present on leaves

❖ Stems thin and few

❖ Poor development of buds

❖ Flowers small and light coloured

Phosphorus deficiency

➢ Leaves dark green.

➢ Leaves may drop before turning yellow.

➢ Stems weak

➢ Poor root development

Potassium deficiency

❖ Marginal browning and scorching of lower leaves

❖ Stems weak and growth stunted

❖ Buds fail to open and flower colour is pale

❖ Short flowering starts with smaller flowers

Magnesium deficiency

➢ Interveinal chlorosis of older leaves

➢ Reduced number of stems

➢ Flowers smaller and poor in colour

➢ Reduced root growth and flower yield

Calcium deficiency

❖ Hard and stubby growth of plant

- ❖ Root system short and thickened
- ❖ Terminal buds die
- ❖ Defoliation of leaves
- ❖ Margins of leaflets die and rest of the leaf turns yellowish green
- ❖ Dark brown patches appear between midrib and margin
- ❖ Leaves drop off before they are dead
- ❖ Flowers deformed, reduced in number and have shown spots near margin of petals

Iron deficiency

- ➢ Interveinal chlorosis of terminal leaves.
- ➢ Younger leaves more chlorotic and transparent

Boron deficiency

- ❖ Terminal buds die and hence excessive branching
- ❖ Leaves crowded together at the tips of the stems giving a tufted appearance
- ❖ Flowers bull headed or malformed
- ❖ Acute deficiency causes death of stem tips

Copper deficiency

- ➢ Plants dwarf and bushy
- ➢ Leaves narrower and twisted.
- ➢ Leaf tips become white

Manganese deficiency

- ❖ Interveinal chlorosis
- ❖ Deformities in tissue formation

Molybdenum deficiency

- ➢ Shortened plant growth
- ➢ Leaves pale and wither
- ➢ Symptoms resemble those of nitrogen deficiency

Sulphur deficiency

- ❖ General yellowing of foliage

2. JASMINE

Iron deficiency

- ➢ Leaves exhibit interveinal chlorosis

- ➢ Leaf size reduced
- ➢ Yield of flowers reduced
- ➢ Rusty brown or red colouration seen on the petals
- ➢ Plants susceptible to fungal diseases at calcareous soil

Remarks

- ❖ Lime induced iron chlorosis is common in calcareous soils. When free calcium carbonate content is high in the soil, uptake of iron by plants is very low.

Remedy

- ➢ Soil application of 50 g Ferrous sulphate mixed with 15 kg FYM per plant in a circular band 60 to 150 cm away form the central stem will alleviate iron chlorosis. Foliar spraying of 0.5% Ferrous sulphate at monthly intervals is also recommended

Boron deficiency

- ❖ Young and terminal leaves become brittle and curled downwards with their tips pale green and curling upwards
- ❖ Petals fail to unroll or reflex properly

3. CHRYSANTHEMUM

Nitrogen deficiency

- ➢ Reduction in plant growth with yellow green foliage
- ➢ Size of younger leaves reduced and plants appear erect
- ➢ Flowering delayed at the initial stage of deficiency
- ➢ In extreme cases, leaf margins become yellow, reddish purple tints develop
- ➢ Leaves die prematurely

Phosphorus deficiency

- ❖ Reduction in leaf size with leaf remaining green
- ❖ New leaves become dark green and lower leaves may develop an orange green colour
- ❖ Marginal browning of leaves, death of affected leaves
- ❖ Flowering delayed with reduction in size and number of flowers

Potassium deficiency

- ➢ Marginal browning and reduction in size of leaves

- ➢ Flowering delayed
- ➢ Keeping quality of cut flowers impaired
- ➢ Glazing and bronzing of leaf underside

Magnesium deficiency

- ❖ Loss of leaf colouration with veins and margins green
- ❖ Older leaves reddish purple in colour
- ❖ Reduced flower size

Calcium deficiency

- ➢ Even slight deficiency causes a complete cessation of top growth and death of roots
- ➢ Roots become dark and adhere together
- ➢ At swae deficiency, yellowing of leaves and internodal shortening
- ➢ Upper leaves form a rosette
- ➢ Buds of growing point die
- ➢ Lower leaves dull dark green, stiff and brittle

Iron deficiency

- ❖ Interveinal yellowing and reduction of leaves
- ❖ Delayed flowering with pale petal colour
- ❖ Plants are prone to fungal rot

Boron deficiency

- ➢ Leaves brittle and down curled with tips pale green and curling upwards
- ➢ Petals fail to unroll or reflex properly

Copper deficiency

- ❖ Leaves pale green, lateral branching excessive
- ❖ Delayed in flowering
- ❖ In extreme cases, interveinal chlorosis and failure of buds to develop

Manganese deficiency

- ➢ General chlorosis and small necrotic spots near leaf edges & petiole
- ➢ Upper leaves reduced in size

4. CROSSANDRA

Iron deficiency

- ❖ Symptoms noticed from the nursery stage itself
- ❖ Upper leaves become chlorotic
- ❖ Veins show thick green colour
- ❖ New leaves smaller and flowering completely affected
- ❖ Spike length reduced; hence flower production greatly reduced with poor quality
- ❖ Length of corollary tube markedly reduced
- ❖ Smaller flowers with short tube and hence marketability is a problem

5. CARNATION

Nitrogen deficiency

- ➢ Limited growth of the plants, thin stems and stiff leaves, which curve at the tip particularly during winter
- ➢ General yellowing of mature or lower leaves
- ➢ Successive leaf pairs develop more slowly and more leaves are formed below the flower so that flower initiation is delayed. Calyx splitting is also often observed

Phosphorus deficiency

- ❖ Restricts the growth of plants and leaves become narrower
- ❖ Tips of lower leaves die off and turn yellow particularly during second year in the field
- ❖ In severe cases, the whole leaf turns yellow or pale straw colour and dies off prematurely

Potassium deficiency

- ➢ Characterized by scorching of leaf blades
- ➢ In severe cases, development of white spots on foliage just below the flower, disfigured or malformed flowers, loss of petal colour and yellowing of calyx

Magnesium deficiency

- ❖ Weak stems and small flowers

Calcium deficiency

- ➢ Production of many sleepy flowers (not opening properly) and consistent foliage scorching

Boron deficiency

- ❖ Pale yellow zone along the midrib, which later turn into purple
- ❖ Reduction in internodal length of flower stem, failure of bud to reach anthesis stage and development of purple colour in leaves
- ❖ Reduced flower production and average bloom diameter
- ❖ Increased incidence of calyx splitting

I. PLANTATION CROPS

1. COFFEE

Nitrogen deficiency

- ➢ Smaller leaves, leaves present a whitish colour; defoliation is common
- ➢ Death of fruit bearing branches takes place

Phosphorus deficiency

- ➢ Irregular, yellowish spots appear over entire leaf blade; chlorotic spots may have a red or purplish tint
- ➢ In acute conditions, these spots coalesce and leaf readily falls off the branch

Potassium deficiency

- ➢ Small brownish spots near the spatial margin
- ➢ The spots coalesce and a dark brown, almost black, necrotic area is formed at the tip of leaf on the adjacent margins
- ➢ Leaf abscises easily

Magnesium deficiency

- ❖ Chlorosis occurs between the main vein and secondary vein
- ❖ A narrow band of green tissue exists
- ❖ Chlorotic spots gradually expand towards the edges of the leaf
- ❖ In acute conditions, a dark necrotic spot appears at the tip of the leaf

Sulphur deficiency

- ➢ Blade of the younger leaves shows yellowish green colour
- ➢ Chlorosis appears as a band, which comprises the main vein and extends towards the middle of the blade
- ➢ Lower surface of the leaf is much lighter in color

Calcium deficiency

- ❖ Younger leaves show chlorotic symptom on their margins
- ❖ Sides of the central vein, the tissue remains green
- ❖ Cupping appears as a result of unequal development of the veins and leaf blade
- ❖ Tissue cracks and suberous tissue is formed
- ❖ Veins assume dark colour due to necrosis in the phloem
- ❖ The apical bud dies under acute deficient condition
- ❖ Yellow areas develop between the veins spreading from the margins
- ❖ Colour of the patches changes to bright orange

Iron deficiency

- ➢ Frequent in the nurseries and in older trees
- ➢ Mild shortage, leaf blade becomes pale green and veins retain the normal dark green colour
- ➢ In acid soil, manganese often builds up toxic levels, thus preventing the adequate uptake of iron

Zinc deficiency

- ❖ Common on acid soils subjected to heavy leaching and erosion losses
- ❖ New leaves are small, narrow and distorted
- ❖ Vein form a green net work against a pale background of yellowish tissue
- ❖ In acute condition, young leaves easily break and have a leather-like appearance
- ❖ Reduced number of leaves. Most of the leaves might fall
- ❖ Shortening of the internodes, a rosette of small, narrow, pointed leaves develop
- ❖ Die-back of the branches sometimes occurs

Boron deficiency

- ➢ Deficiency is death of terminal buds; and such dead branches and top of the plant remain attached to the plant for some time
- ➢ Few branches develop below the dead terminal bud giving the appearance of a fan
- ➢ Leaves are usually narrow and twisted with irregular edges; internodes are shortened

2. TEA

Zinc deficiency

- ❖ Considerable loss in leaf yield and decline in quality of tea
- ❖ Internodes are short compared to normal plants
- ❖ Leaf size highly reduced
- ❖ Yellowish chlorotic spots; rosett- ing of leaves, sickling of leaves and wavy margin in leaves are noticed
- ❖ Tea bushes become stunted
- ❖ In acute deficiency, defoliation takes place, reducing the vitality of the plant

Remarks

- ➢ High organic matter content in the soil which fixes zinc as zinc humate, continuous use of high analysis fertilizers, low content of zinc in the soil and accumulation of zinc at collar region will lead to zinc deficiency

Sulphur deficiency

- ❖ Lamina of the younger leaves shows yellowish green colour
- ❖ Chlorosis appears as a band, which comprises the main vein and extends towards the middle of the blade
- ❖ Lower surface of the leaf is much lighter in colour than the upper one
- ❖ Tea is an indicator plant for S deficiency causing typical Tea yellowing of leaves

Manganese deficiency

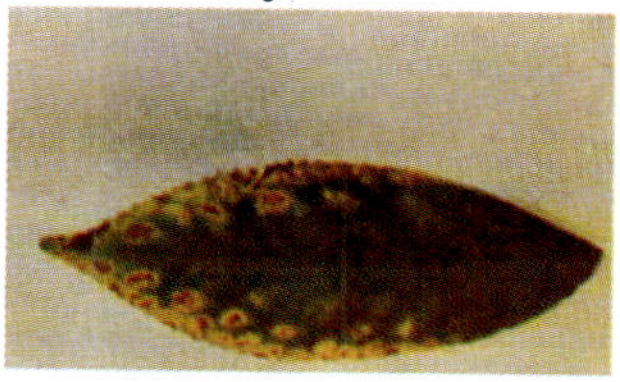

- ➢ Show chlorosis or light green between the veins.
- ➢ Small mottles, yellow or dark brown in colour are found on the edges of the leaves.

3. CASHEW

Nitrogen deficiency

- ❖ Paling of green colour of the older leaves, which later on change to uniform yellow

Phosphorus deficiency

- ➢ Normal green colour of the leaves changes to dark green
- ➢ Withering of lower leaves
- ➢ Under severe deficient conditi-ons, the leaves take purple colou-ration
- ➢ Reduction in plant height and leaf number

Potassium deficiency

- ❖ Yellowing of lower leaves and necrosis of older leaf tips
- ❖ Height of the seedling and leaf production also greatly affected

Magnesium deficiency

- ➢ Interveinal chlorosis and yellowing of lower leaves are the charact-eristic visible symptoms

Sulphur deficiency

- ❖ Deficiency appears as pale green to greenish yellow discolouration of younger leaves, which later turn to yellow
- ❖ Small necrotic spots appear on the affected leaves, which coalesce to become larger necrotic areas
- ❖ Shedding of affected leaves and growth retardation

Iron deficiency

- ➢ Observed on the calcareous soils of cashew plantations
- ➢ Chlorosis appears on the younger leaves; yellowing of immature leaves
- ➢ In addition to the interveinal chlorosis, reduction in growth rate and premature leaf fall

Zinc deficiency

- ❖ Interveinal chlorosis of younger leaves
- ❖ New leaves produced are smaller in size, terminal growth is retarded and internodal length diminished resulting in rosette conditions
- ❖ Abscission of leaves

Manganese deficiency

- ➢ Observed on the seedlings of cashew and it is known as Grey spec disease
- ➢ Younger leaves develop interveinal chlorosis; chlorotic patches turn necrotic and reddish brown
- ➢ Leaves dry and fall off
- ➢ Young emerging leaves show a typical symptom called mouse-ear, wherein leaves show dark brown interveinal mottling and the leaves droop down and become brittle
- ➢ Brownish patches near leaf margins; leaf development will also be poor
- ➢ Chlorosis may cover the entire lamina making the green major veins and laterals more conspicuous
- ➢ Dwarfing and reduced leaf production
- ➢ Chlorotic leaf areas become necrotic and turn reddish brown and the leaf dries and falls off

J. NARCOTICS

TOBACCO

Nitrogen deficiency

- ❖ General growth of plant poor
- ❖ Smaller leaves
- ❖ Lower leaves turn yellow

Phosphorous deficiency

- ➢ Leaves dark green or brown
- ➢ Short stem

Potassium deficiency

- ❖ Thin stem and blue green leaves with necrotic margins and tissues

Magnesium deficiency

- ➢ Discolouration between veins beginning with lower leaves
- ➢ Leaves present a crumbly appearance

Zinc deficiency

- ❖ Leaves thickened; symptom expressed more strongly in summer

Copper deficiency

- ➢ Loss of turgidity in young leaves and stems

Manganese deficiency

- ❖ Marbled yellow upper leaves beginning from the blade tip; veins remain green.

4

PHYSIOLOGICAL DISORDERS

It is common knowledge that a crop plant is in need of all the 16 essential nutrient elements for its proper growth and development. If the plant is not supplied with any one or more of the essential elements, the growth of the plant will not be normal. As and when the plant experiences shortage of any nutrient element at any stage of its life cycle, the plant will put forth externally various visual symptoms of deficiency. The deficiency of a nutrient element may be manifested in the form of either one or more of the following visual symptoms: chlorosis, necrosis, die-back, scorching, browning, reddening, yellowing, general stunted growth, etc.

Not only does deficiency of a particular nutrient adversely affect the growth of a plant, but also some of the physiological disorders are known to have unfavourable impact over the growth and development of the plant. These physiological disorders may arise as a result of several factors such as defective soil physical and chemical properties, unsuitable irrigation water, sudden shift in the various environmental factors and undesirable cultural practices. Among these factors, those relating to environment play a significant role in bringing about several physiological disorders in both agricultural and horticultural crop plants and they include light intensity, excessive moisture, drought, high wind velocity, heavy rains, lightning, extreme ranges of temperature and humidity.

Diagnosing physiological disorders in crop plants

Besides nutrient deficiency problems, farmers also face problems brought on by factors other than nutrient deficiencies. Some problems are caused by pests, others by diseases and yet others are caused by a variety of environment induced factors. These disorders are also called physiological or environmental disorders. Some physiological disorders are easy to diagnose, while others are difficult or even impossible to diagnose. Identifying the causes of physiological disorders is like solving a puzzle and challenges farmers to stretch their imaginations, make keen observations and put it all together with the help of their experience. Some physiological disorders are beyond help once symptoms are visible. Nevertheless, it is extremely important to arrive at the right diagnosis in order to avoid the problem in the future. It is important for growers to know that:

- ❖ Physiological disorders are often caused by lack or excess of something that supports life (e.g. lack or excess of micro-nutrients) or by the presence of something that interferes with life (e.g. a pollutant).
- ❖ Physiological disorders occur in the absence of pathogens and cannot be transmitted to healthy plants. Different plant reactions to the same disorder vary widely from asymptomatic to severe, near - death damage.
- ❖ The symptoms vary in kind and severity depending on how much of the causal factor deviates from its norm: a small concentration of ethylene may not be noticeable while a concentration slightly larger produces noticeable symptoms.
- ❖ Dealing with physiological disorders often means dealing with the consequences of something that happened in the past.
- ❖ Physiological disorders are not only serious problems in and of themselves but also may serve as the "open door of entry" for pathogens.

What to do when facing physiological disorders?

The cure, if possible once symptoms are visible, requires:

1. Removing or neutralizing the causing agent from the plant's growing environment or
2. Supplying the substances whose lack caused the problem.

Identifying the cause of a physiological disorder can take time, effort and patience. It is not uncommon to find disorders whose cause cannot be identified. The impossibility of reconstructing past events or high cost of tests or analyses can make identification very difficult.

Some examples of factors causing physiological disorders follow:

Water

Water can be the source of numerous problems, especially in floriculture. Water related problems fall in three categories:

- ➢ Too much water
- ➢ Too little water
- ➢ Problems caused by poor water quality (mineral content)

Too much water is very common in container grown plants. Too much water leads to a lack of oxygen; anaerobic microorganisms (those which grow without oxygen) produce nitrites that are toxic to roots; roots lose selective permeability.

How is this type of stress recognizable?

Over-watered plants:

- ❖ drop lower leaves suddenly
- ❖ have leaves that turn yellow
- ❖ show black wet patches
- ❖ roots or lower stem may turn black
- ❖ may have succumbed to a disease
- ❖ show wilting

The problem with this list is that many other diseases and/or disorders can cause similar symptoms. Ultimately, questioning the grower and finding out what happened, for example, with the irrigation system will lead towards the cause.

Too little water is even more of a problem on a global scale and very common with container grown plants. Water stress can be categorized in four different subcategories as shown below.

Single Severe Stress	Periodic stress	Continual stress	Low to High Stress
• Tip burn of mature leaves	• Poor growth	• Small leaves	• Wilting
• Young leaves are dead, crispy but green	• Dark green (similar to high salinity)	• "Burned" old leaves	• Leaf drop
• Young leaves are distorted when plant recovers	• Some plants have defoliation	If no rain or watering the plant dies	• Damage in leaves; young leaves black
• This is referred as weekend syndrome	• Smaller plants than normal	(High EC can cause similar symptoms)	

As with excess water stress, these symptoms are, for the most part, common to other problems and ultimately they have to complement with more evidence obtained by questioning the growers, studying environmental records, or inspecting irrigation systems.

Poor water quality is a common problem in arid areas of the country and with crops grown in containers. The water characteristic that cause problems are:

- electrical conductivity (>1.5 mS)
- alkalinity (>120 ppm)
- too much of some element (e.g. Boron, Sodium)

Temperature

Each species/cultivar has an optimum range of temperatures. Exposure to extreme from that optimum causes problems. These problems are more severe, the farther temperature deviates from the optimum. Temperature problems can be divided into either being too high or too low:

Temperature too high is usually associated with high light, low relative humidity, and/or insufficient water.

Usually high temperature and lack of water have similar symptoms: dead dry tissue that remains on the plant that can be easily broken (it does not bend). Usually young tissue is affected earlier than old tissue.

High temperature affects plants by inactivating certain enzyme systems and accelerating others; or by coagulating proteins. Depending on the species, low-temperature damage can be caused by chilling temperature or frost. The injury varies depending on the degree of temperature drop or the duration of the low temperature period. Low temperatures injure plants by inducing ice formation between or within the cells.

Low Oxygen (in the soil)

Low Oxygen is a consequence of water logging. It can be aggravated by high temperatures. Low oxygen also occurs in the interior of fruits and tubers when the temperature is high. This situation is more common in storage.

Light

Lack of light produces etiolated (pigment deficient) plants. Stretching can be the consequence of low light levels. Excess light is rare but may occur with indoor plants exposed to sun light or plants grown in the shade and suddenly exposed to sun.

Air pollutants

Most pollutants are gases. Ethylene, ammonia, chlorine, and mercury vapours are all possible problems in closed areas like greenhouses and storage facilities. Sulphur dioxide, ozone, fluorides, and peroxyacetil nitrate are associated with "smog" and can cause problems with plants outdoors. Un-vented heaters and problems with the heater's exhaust are very common in the greenhouse industry. These problems are more noticeable in poly houses (because they are more air tight) than in glass houses.

Symptoms of fume injury

- Top growth deformed
- Stunned growth
- Epynasthy (petioles and leaves bend downward)
- Flower bud abortion in some crops (e.g. mums)

Gas injury

Plants can be injured and killed by gases in the soil. Gases may be directly toxic or reduce oxygen to critically low levels.

- Natural gas is not toxic *per se*.
- Manufactured gas contains a number of toxic compounds.
- Gases produced by the decomposition of organic matter in the landfills can be toxic to plants in certain concentrations.

Herbicide Injury

Herbicide injury can occur either when herbicides are applied in the greenhouses where the crops are located or outside the greenhouses (green space between greenhouses or crops in the vicinity). Herbicide injury can be caused by:

- Direct application to the crop
- Persistence of the herbicide in the soil affecting the following crop
- Volatilization in closed environments

How to approach these problems

The first step is to do an exhaustive collection of data. If the information collected does not give clear indications of non-infectious disorders, then test for pest or diseases. If pests and diseases can be ruled out, test for fertility problems. After ruling out fertility problems, start considering a more complicated noninfectious disorder. The following steps need to be taken

- Make observations
- Describe what is wrong with the crop
- Ask appropriate questions
- Look for patterns
- Spreading in space
- Spreading in time

Test for pests and diseases

- Virus

- Insect
- Bacteria
- Mites
- Fungus
- Others

Test for fertility problems

- Nutrient deficiency
- Nutrient toxicity
- Water quality

Other reasons
(Non infectious disorders)

- Pollutants
- Cultural practices

Collection of information

If farmers are not able to find the problem by themselves after going through these questions, they will have to discuss the problem with their Extension Officer or subject matter specialist.

1. Describe the plant

- What type of plant is it (genus, species, cultivar)?
- How old is the plant; when was it planted?
- Did you grow it or purchase it from someone else?
- If purchased, who did you buy it from (clues to the plant's culture)?
- Did you purchase cuttings, seed or seedlings, or a potted plant?

2. Describe the plant symptoms

- Are there any obvious symptoms of insects or disease such as webs, chewed leaves, honeydew, larvae, shed skins, mold, rust, blackened areas on stems or leaves, root rot, or wilting?
- Are the symptoms restricted to certain varieties?
- Does the problem occur in particular locations in the bench or in the green-house, or is it scattered throughout the crop? Is the problem mild or severe?
- How was the crop grown?

3. Outline the case history

- On what date did you first observe symptoms?
- Did the symptoms appear all of a sudden or gradually?
- Did you apply any treatments to the problem?
- Did they work?
- Have you had previous problems of any kind with this crop?
- Is what you see now a recurring problem?

4. Describe the plant environment

- Is the plant in a container or planted in a bench?
- Are the benches raised or on the ground?
- What are the location, age, and condition of greenhouse benches?
- What is the potting or bench mix used? Did you make it or buy it?

- ❖ Was it pasteurized?
- ❖ How was it pasteurized (steam, chemicals)?
- ❖ Did the mix contain any starter fertilizer?
- ❖ What kinds and how much?
- ❖ What were the day and night temperatures, light level, and humidity in the greenhouse?
- ❖ What chemicals were used during crop production - fertilizer (types, amounts) - pest controls (sprays, fumigants, granular materials, etc.) - growth regulators - weed control (herbicides)?
- ❖ Were label instructions followed carefully?
- ❖ Did you use spreader stickers?
- ❖ If so, which ones?
- ❖ What is the age and conditions of equipment you use in crop production?
- ❖ How accurate is the fertilizer injection system?
- ❖ When it was last calibrated?
- ❖ How about the heating system and the water supply?
- ❖ What is the type, age, and condition of greenhouse covering?
- ❖ Are there any septic tanks or gas lines near the greenhouse which could have caused air pollution problems?
- ❖ Have herbicides been applied around the greenhouse or under the benches?

5. Consider the general details of plant culture as well as any special events

A failed mechanical system or an improperly executed operation that may have caused a problem.

- ➢ Can the symptoms be associated with any particular age or stage of crop development?
- ➢ Can the symptoms be associated with weather conditions or season?
- ➢ Do the symptoms get worse over time or stay the same once established?
- ➢ Describe in detail the condition of the roots, stems, foliage, and flowers.

The various physiological disorders commonly encountered in some of the horticultural tree crops are highlighted in this book. The crops covered in this section are mango, banana, citrus, grapes, guava, sapota, pineapple, pomegranate and apple besides coconut. The contributory factors for the onset of various physiological disorders in these crops are dealt with and appropriate remedial measures indicated to alleviate the disorders

1. MANGO

Malformation in mango

Mango malformation is a serious physiological disorder causing great concern and heavy losses. There are two types of malformation noticed in mango:

(a) Vegetative malformation

(b) Floral malformation

(a) Vegetative malformation

It is more common in the nursery seedlings and young plants, wherein malformed vegetative shoots are characterized by the appearance of small leaves in whorls on thick, stunted shoots which may crown the apex, giving a bunchy-top like appearance

Remedy

Spray of 0.5 % $ZnSO_4$ +1% Urea will alleviate the disorder.

(b) Floral malformation

Of these two types of malformation, floral malformation causes heavy losses in fruit yield. In this case, the floral parts are transformed in to compact masses of sterile flowers appearing in the form of a bunch, causing total loss to the crop.

It is characterised by the deformation of panicles, suppression of apical dominance, shortened primary and secondary axis, thickened rachis of panicles (thus giving the flowers characteristically clustered appearance) & a preponderance of staminate and large flowers. Such panicles rarely set fruits, ultimately dry up and persist on the tree for a longer time

Causes for floral malformation

Higher levels of GA_3, ABA (Abscissic Acid) and lower levels of IAA (Indole 3 Acetic Acid) are responsible for increased rate of production of floral malformation in mango.

Physiological and biochemical changes in malformed floral buds

Various causes such as cultural practices, nutritional practices, mites, virus, fungi, etc. have been attributed, but the precise causal organism / substance is yet to be established. Nevertheless, following physiological /biochemical changes are found to occur in the malformed buds:

(a) Level of C/N ratio in malformed buds is higher, which is responsible for the development of more staminate flowers in malformed panicles, suppression of flower development and fruit set.

(b) Carbohydrate contents are invariably higher in shoots bearing malformed panicles. Starch did not hydrolyze into sugar to meet their energy requirement

(c) Total protein content is lower because of lower nitrogen content

(d) Micronutrient deficiencies of Fe and Zn are often associated with the malformation

(e) Malformed panicles contain lower nitrogen content enhanced application of N curtails malformation

(f) Auxin (IAA) is lower with increased contents of GA_3 and ABA.

Correction

One spray of 200 ppm of NAA (@ 20 ml of Planofix dissolved in 4.5 lit. of water) during first week of October is effective in reducing the incidence of malformation. The malformed panicles must be pruned in the month of April and burnt.

Biennial bearing in mango

The problem of biennial bearing (tree carries optimum load of crop in one year but in the following year it fails to flower and produced unsatisfactory crop) is known to cause tremendous loss to the mango industry in the country.

Control

Use of ethylene releasing compound, Ethrel (Ethephon) @ 200 ppm coupled with 0.1% urea is found to control alternate bearing and also to induce flowers every year. In all, five sprays are to be made beginning from middle of September at monthly interval.

Soft nose in mango

1. The interactive effect of nitrogen and calcium in the soil causes this disorder.
2. In acid sandy soil, the incidence is about 7% only in trees receiving low levels of nitrogenous fertilizer, while it is increased to 78% in trees receiving 10 times more of N fertilizer.
3. When calcium is kept at 2.5% or slightly higher, the incidence of soft nose gets reduced appreciably.

Control Measure

It is suggested that calcium content of the soil should be maintained by the application of N in the form of calcium nitrate of gypsum (@ 1kg/tree) or lime stone.

Black tip in mango

In India, mango necrosis or black tip is common.

Yellow spots first appear at the distal end of the fruits, which later on turn brown and finally black. This disorder usually occurs in orchards in the vicinity of brick kilns. The affected fruits become unmarketable and as such contribute to the economic loss to the growers. The loss may even touch a total loss of the crop depending upon the number of operating brick kilns and their distances from the orchards. The loss due to black tip is very serious in orchards located within 3-4 km from the brick-kiln but during very dry weather conditions (April-May), the damage is more severe and also noticed in the orchards located far away (up to 8 km) from the brick-kilns.

The fumes that come out of the brick kilns as a result of burning of coal contain carbon monoxide, ethylene and sulphur dioxide. All these gases are toxic to the young developing fruits of mango but the acidic fumes of SO_2 are the most harmful in accentuating this disorder. There is also interaction between boron content of the plant and fumes of brick kiln causing the disorder.

Correction

This disorder can be minimized by the spray of 0.6 % borax and 0.8% caustic soda thrice, *ie.*, before flowering, during flowering and at the fruit-set stage. Some sticker or washing soap may be added in the spray solution. It is emphasized that the stage of application is very important.

Fruit drop in mango

The problem of fruit drop is quite serious in some of the commercial varieties of mango and cause great loss to the growers.

A tree produces several thousand panicles but yields only a few hundred fruits, most of the flowers falling down after full bloom or at later stages of development.

Fruit drop at the initial stage (pin head drop) may not affect the yield much, while the same at later stages (post setting stage, May drop phase) drastically reduces the yield.

Causes of fruit drop

The following may be the reasons for the fruit drop:

1. Abortion of embryo
2. Degeneration of ovules
3. Depletion of nutrients
4. Hormonal imbalance

Correction

Fruit drop may be controlled by sprays of NAA or 2,4-D (@20 ppm) at pea size of the fruit.

Spongy tissue in mango

This disorder is common in the fruits of Alphonso wherein a non-edible, sour, yellowish sponge-like patch with or without air pockets develop in the mesocarp of the affected fruits during ripening. Externally, the affected fruits present a healthy appearance.

The incidence is reported to be more if fruits are harvested at full maturity than 3/4th maturity. Exact cause for the malady is not known as on today. However, excess application of N would probably cause the disorder.

Remedy

Regulated supply of N and enhancing the concentration of Ca in fruit tissue by the soil application of gypsum @ 1 kg/ha.

Scorching of leaves

It has become a serious problem in north India. The typical symptoms are the brick red color towards the tip, along the margins of old leaves and subsequent collapsing of these tissues. New leaves do not exhibit any such deformities. Almost all leaves get affected in the winter. The symptom resembles potassium deficiency, but it is caused by chloride ion toxicity. For K application, the use of potassium chloride should be avoided, instead potassium sulphate should be used. Irrigation water having high chloride content should not be used.

2. BANANA

Seediness in banana (kottai-vazhai in poovan)

Poovan (AAB) is by far the leading commercial variety of banana occupying morethan 70% of the area under banana cultivation in India and manifests a peculiar development disorder, a predisposition to seediness, locally known as "Kottaivazhai".

Characterization of disorder

This malady is characterized by the presence of distinctly conical and ill-filled fruits with a prominent central core having many under-developed seedy structures rendering the fruits inedible. It has a potential to reduce the production by 10-25%. Because of the deceptive nature, this disorder poses a serious threat to engulf the entire area under Poovan imperceptibly and the banana growers are getting bewildered. This malady could be identified at the later stages of development from the appearance of streaks, striations and blotches on the pseudo stems and on the leaves.

This is also genetically inherited, as there is a presence of heavy and intense blotches and striations even on the second crop of "Kottaivazhai". The bunches are held at an angle above the horizontal position. The peduncle showed irregular curvature and exhibited a shape.

Pollen of Kottaivazhai is infertile, shriveled, shrunken and broken. The seeds are soft, flat, ill -filled and regularly angled.

There is not much difference in the basic structure of the fruits of normal Poovan and Kottaivazhai. The area of the pericarp in Kottaivazhai is much smaller than that of normal poovan. The locular cavity of Kottaivazhai is bigger than that of normal Poovan fruits.

Remedy

Spray of both Gibberellic Acid (50 ppm) and 2,4-D (20 ppm) when made immediately after the last hand is opened, the seediness in Kottaivazhai is completely overcome and the treated bunches is found better than normal Poovan in respect of the bunch and fruit characters and quality.

Goose flesh in banana

The symptoms of this disorder observed on ripe fruits and the degree of damage varies from slight to severe. Affected fruit has a wilted and shriveled appearance and peel usually turns brown, when the fruit is handled.

Causes of disorder

This disorder occurs more due to dry winter climate, when the atmospheric humidity falls below a low level.

Goose flesh symptoms are caused by disruption of the cellular structure, which is a result of partial dehydration of the peel during ripening. Ripening of banana fruit is characterized by conversion of starch to soluble carbohydrates. This reaction increases the osmotic pressure and draws moisture from the peel to the pulp.

Remedy

Proper storage conditions and careful handling of fruits would reduce the incidence of this disorder. Increasing the moisture of the peel by spraying 0.5% KCl during early stage of fruit development would be rewarding.

Comb-teeth effect in banana

Cause of Disorder

Occurrence of this disorder is due to the manganese deficiency in the orchard.

Symptom Characterisation

In this disorder, chlorosis is marginal at first, sometimes with a tin green edge remaining on the leaf margin. It spreads towards the midrib but more quickly to the main veins than to the interveinal areas, which remain greener from a long time giving a "Comb Effect". These symptoms affect the entire plant, its foliage turning a dirty yellowish green.

Degrain of bunch in banana

This disorder is characterized by the dropping of ripe banana fruits from a bunch due to rotting of the pedicels. Long stalks are mechanically weaker than the shorter ones and they also differ anatomically. High N/P ratio in the manure tends to thicken the stalks and flatten the central cylinder and markedly increases the susceptibility to tannin oxidation.

Remedy

Enhanced application of phosphorus (preferably DAP application) would alleviate the disorder.

Finger drop in banana

This disorder is often related to climatic conditions favouring excessive photosynthetic activity especially during post-flowering period, when the plant is unable to exert a modifying effect on any sudden variations in the soil N supply. Abundant K supply probably reduces the disorder.

Yellow pulp or green ripe disorder

This disorder is characterized by premature and abnormal development of immature banana fruits.

Cause

Excess application of K in relation to N is attributed to this malady.

Correction

Regulated supply of K in conjunction with the N application.

Choke throat

It is due to low temperature affecting active growth of the plant. Leaves become yellow and in severe cases, the tissue gets killed. In case of normally flowering plants, the stalk carrying bunches elongates freely so that the entire inflorescence comes out of the pseudostem and hangs down. Bunch development is normal, but when the time of flowering synchronizes with low temperature, the bunch is unable to emerge from the pseudostem properly. The distal part of the inflorescence comes out and the basal part gets stuck up at the throat. Hence, it is called *Choke throat*. Maturity of the bunch is delayed by taking 5-6 months instead of 3. 5- 4 months for harvest.

Precautions provision of shelter belts, planting low temperature tolerant varieties, adjusting the time of planting and orchard to prevent rapid cooling during cold spells.

Chilling injury

Chilling occurs when pre-harvest or post-harvest temperatures fall below 14°C for various time periods. The peel of banana become dark and the fruit exhibit uneven ripening. Ripening fingers show dull yellow to smoky yellow colour and watery dark patches on the skin. Brittleness of the fruit and fungal invasion is also observed. The vascular bundles of the sub epidermal layer show brown streaks. The discoloration is ascribed to the enzymatic oxidation of dihydroxy phenylalanine.

Hard lump

Hard lump or masses of varying sizes and shapes occur in the pulp. Lumps are pinkish brown in colour and more firm than the usual pulp and taste like immature or unripe fruits, with the result, the quality of the fruit is affected. Occurrence of hard lump is seasonal and only the bunches shot during dry winter period exhibit the phenomenon. The pink discoloration and astringent taste of fruit is due to high accumulation of tannin during winter months, which inhibit the conversion of starch to sugar. Spraying or dipping the end of peduncle in 20ml of 2,4-D 1000 ppm checks the disorder.

3. GRAPES

Coulure in grapes

Coulure (French word meaning shelling) is a physiological disorder referring to the premature drop of bud, flower and /or berries in grapes.

Grape growers are confronted with the problem of bud, flower and berry drop. This disorder is found severe in the varieties of seedless type (eg. Thompson Seedless, Pusa Seedless, Beauty Seedless, etc.). In many of these cultivars, unopened flower buds and flowers drop heavily and result in poor fruit set. The loss due to coulure has been estimated to be 70-80%. This disorder causes a characteristic partial or complete barrenness in the veins.

Factors affecting coulure

1. Carbon Nitrogen Ratio
2. Plant Growth Regulators (PGRs)
3. Development Process and Nutrition
4. Climatic Factors

1. Carbon / Nitrogen (C/N) Ratio

Proper C/N ratio in the bearing shoots has been reported to be associated with fruitlessness. Excessive manuring leads to an imbalance in C/N ratio resulting in Coulure, insufficient bud differentiation and bud killing.

The disorder also seems to be associated with low C/N ratio in grape vine. Cultivars having high C/N ratio (like Champion, Perlette etc.) are generally free from bud, flower or berry drop, whereas seedless varieties having C/N ratio of less than 6.0 are found to be susceptible to this disorder.

Girdling causes uniform fruit ripening and increased C/N ratio and pedicel thickness. This has happened by altering the distribution of photosynthates or carbohydrate level in the vine and thereby checking the Coulure disorder.

2. Plant Growth Regulators (PGRs)

Levels of endogenous growth hormones may affect bud, flower and berry drop.

(A) Auxins

Applied auxin maintains on-going physiological and biochemical functions. In grapes, endogenous auxin level is found to be low preceding the young fruit drop and the pre-harvest drop, which could be checked by the exogenous application of auxin. Application of auxin has been found to prevent the synthesis of cell wall degrading enzyme, cellulase. Therefore, application of auxin often results in checking abscission.

(B) Cytokinins

Cytokinins are believed to maintain the abscission zone in juvenile state and thus prevent abscission of flowers, buds or fruits. Benzyl adenine might be effective in decreasing berry drop by delaying senescence. It has been found increased pedicellar attachment as a result of kinetin application.

(C) Growth Retardants

Partial or complete cessation of vegetative growth creates a condition of increased carbohydrate content and thereby forms a good fruit set. Thus, growth retardants like Cycocel (CCC: Chloro Choline Chloride) are found to increase fruit set in several grape cultivars. Such beneficial effects have been attributed to the increased biosynthesis of cytokinins through CCC application.

Thus, a condition of low level of endogenous growth promotors like auxins, gibberellins and high level of growth inhibitors like ABA, leads to hasten the abscission of bud, flower and berry drop and poor set in grapes..

3. Development process and nutrition

Failure of fruit set occurs in grapes due to poor pollination and improper fertilization owing to unilateral compatibility, imperfect flowers, poor pollen germination and fertility or ovule abortion.

Since boron is responsible for the normal development of fruits, its deficiency might result in poor set. Spray of sodium borate (@ 0.01%) or boric acid (@ 0.2%) at full bloom increases fruit set in grapes.

Shattering of flowers and set of straggly clusters with many shot berries is found to be due to deficiency of zinc. Foliar application of 0.5% $ZnSO_4$ increases the fruit set and yield.

4. Climatic Factors

Temperature affects pollen germination and pollen tube growth and thereby fruit set and flower abscission. At 15.5°C pollen tube growth is very slow and becomes rapid if the temperature rises.

1. Low light intensity results in poor carbohydrate due to reduced photosynthetic activity and thus causes berry abscission. Thus, shading by excessive foliage of cloudy weather causes bud, flower and berry drop.
2. Rainfall during full bloom dilutes or washes away the stigmatic fluid and causes poor pollen germination. Therefore, areas with abundant rainfall often have coulure in grape vines.
3. High relative humidity is found to decrease coulure disorder.
4. Persistent water stress during flowering hastens abscission. Water stress during the first four weeks following bloom has a marked decreasing effect on fruit set. Water stress may affect fruitfulness indirectly through a decline in photosynthesis, by decreasing the cytokinin in the xylem sap or by increasing the level of ABA in the leaves and stems.

Correction measure

1. Optimum and timely cultural operations like irrigation and nutrient application.
2. Application of FYM mineral fertilization combination
3. Judicious adoption of orchard management practices like training, girdling, etc.,
4. Spray of NAA, IBA to prevent preharvest berry drop.

Bunch wilting

It is common in South as well as North India. This disorder specially occurs at ripening stage. The main symptoms are shriveling of rachis and drying of berries at the tip of the bunch. The cause being high temperature coupled with moisture deficit conditions and irregular water relation in the transpiring leaves and developing bunches. The malady can be reduced by proper mobilization of carbohydrate in the vine. Spray of 0.2% boric acid can control by increasing sugar translocation. Pink-berry formation problem of Maharashtra also seems to be due to rising temperature, late in the season.

Blossom-end rot

A black sunken spot develops at the blossom-end of the berry which later on spreads with water-soaked region around it. Defective calcium nutrition and assimilation appear to be the cause for it. Spray of 1.0 per cent calcium nitratet is the corrective measure.

4.COCONUT

Crown choking in coconut

The incidence of this malady is first observed in 1964 in Assam and West Bengal. Young palms of the age group 5-

10 years are mostly affected. Nevertheless, the disorder is noticed in the bearing palms also.

Symptoms

The first symptom is the emergence of shorter leaves with deformed and crinkled leaflets, which are associated with sever tip necrosis. These deformed leaflets fail to unfurl and ultimately give a chocked appearance to the frond. Hence, this deformity of the palms is called Crown Choking.

In case of young palms, peripheral leaves crowd around the bud and prevents normal unfurling of the flag leaf. In acute cases, necrosis of the primordial tissue takes place and the crown dies, but not suddenly. The affected palm loses vitality slowly and succumbs finally within 3 to 4 years.

Tissue Testing

The analysis of leaf samples and soil indicated that the calcium content of the affected palms showed significantly high, while the boron content is in very low concentration.

Control

Application of borax @ 50-100 g/palm to the coconut basin results in the recovery of the affected palm.

5. PINE APPLE

Crook neck disease

It is caused by the physiological disturbances of the plant owing to zinc deficiency. This disorder is attributed to copper deficiency which has an unfavourable influence on the absorption of zinc even when there is adequate reserve of available zinc in the soil. The crook- neck disease generally occurs when copper supply is insufficient.

Remedy

The disorder can be cured by foliar spray with 1% $ZnSO_4$ and spraying in the soil around the plants with 1.5-2.0% $CuSO_4$, avoiding trees.

Sun-scald

Sometimes the peduncle bearing the fruit leans or falls on one side exposing the fruit to direct sun rays. The cells under the skin of exposed surface get damaged. Care should be taken to control lodging or leaning and the fruits should be covered with dry straw or banana leaves or with its own leaves during April-May. Potassium deficiency is also noticed in pineapple with blackened surface at base of the fruit.

6. CITRUS

1. Citrus die back

Nature of the disorder is physiological since no organism is found to be associated with it.

Symptoms

Whole plantations, individual trees, trees in the patches may be affected. The first symptom appears when the terminal leaves and shoots turn yellow. This yellowing is different from simple mottling or the yellowing of senescing leaves. The symptoms appear all at once in the new shoots of fresh flush . The trees may be apparently in perfect health, before the flush, but all of a sudden, they begin to produce pale tender leaves, which do not mature at all. After some time, they attain considerable turgidity and remain stunted and pale. Often variegated leaves are also observed. This variegation is also caused by regular stripes of green and yellow colour alternating one another.

June drop

This disorder is characterized by abnormal drops of flowers and fruits as observed on some groves during certain seasons.

In every grove, some amount of flowers and fruits are bound to drop even under normal conditions and june drop is the term applied to the heavy dropping period which is characteristic in American citrus orchards.

Causes

1. Insufficient supply of N or moisture
2. High wind velocity
3. Low temperature
4. Sudden variation in weather
5. Excessive water stagnation
6. Deep ploughing / other tillage practices causing disturbances to the root at the blooming or early fruit setting
7. Pests and diseases

Correction

Foliar spray of 2,4-D @ 5 -25 ppm reduces the fruit drop considerably.

Fruit splitting / cracking in citrus

It refers to the splitting or cracking of fruits in oranges and certain lemon varieties. It is more severe in Mosambis. Dry period followed by heavy rain and high relative humidity is mostly responsible for this disorder. Cracking is the effect of irregular physiological changes in the plant tissues caused by irregular watering and manuring. Considerable damage is caused to the crop by this trouble, but its severity is limited.

Correction

Abundant supply of K would reduce this disorder.

Granulation in citrus

It refers to drying of the juice vesicles in fruits, resulting in decreased contents of soluble solids acids and increased accumulation of polysaccharides like starch, cellulose and pectin.

Remedy

Foliar sprays of micronutrient (Zn + Cu + K each at 0.25%) reduce the granulation percentage.

Degreening in citrus

In certain varieties of oranges and lemon, fruit colour remains green even after attaining maturity. As such, the marketability of the fruits is reduced. The fruits can be degreened artificially by spraying Ethrel @ 2000 to 4000 ppm at pre-harvest stage.

7. GUAVA

Fatio disease in guava

The affected leaves show red spots, which coalesce and subsequently dry up. The branches develop cracks and finally die back. In extreme cases, breakdown of internal tissues of fruits is also observed.

Cause

Deficiencies of organic matter, nitrogen, zinc and boron may be the reason for the disorder.

Remedy

Application of FYM @ 50 kg /tree and foliar spray of $ZnSO_4$ and 0.2% boric acid would alleviate the disorder.

8. APPLE

Internal bark necrosis in apple

This disorder is also known as Apple Measles or Rough Bark disease, which is a serious problem in all parts of the world.

Symptom

The characteristic symptoms of the disorder is the elevation on the epidermis, pimply lesions, with inner bark necrosis, stunting of growth, die back of shoots and interveinal chlorosis of basal leaves.

Causes

Severity of the disorder increases with increasing levels of K and Mn and with lower levels of Ca necessary for normal growth and development. Higher levels of K enhance Mn toxicity. Lower concentrations of B and P also contribute for the disorder. Recommended dose of macro and micro nutrients would solve the problem.

Bitter pit in apple

Many factors are found to be associated with the disorder. Among them, nutrition seems to be the most important one.

The disorder is characterised by small brown spots in the flesh usually near surface and around calyx of the fruit. Warm weather and water stress during fruit maturity favour the disorder.

Causes

Too early harvesting of fruits, heavy pruning, excess nitrogen and low calcium levels are the probable reasons for the disorder. Also delay in cooling, slow cooling and higher storage temperature favour the incidence of bitter pit. It is proposed that Ca moves out of fruit during dry weather at critically low calcium levels, thereby leading to damage of cells and development of bitter pit.

Remedy

Lesser incidence of the bitter pit is noticed with the uniform supply of N up to harvest. Alternations between high and low N levels aggravate the disorder. Nevertheless, lesser bitter pit incidence is noticed when the trees are supplied with ammoniacal nitrogen in summer and autumn seasons.

Low fruit Ca level is associated with the occurrence of bitter pit. Therefore, the incidence can reliably be predicted by the ratio such as K/Ca, Mg /Ca or (K + Mg / Ca), since higher K and/ or Mg supply promotes bitter pit by lowering the fruit Ca level.

However, the most reliable indicators of susceptibility to bitter pit are :(K + Mg / Ca) and Ca content of fruit at harvest. A threshold (K /Ca) ratio of 12 and a threshold Ca content of 47 mg/100 g dry matter (0.047%) are established.

Application of gypsum @ 1 kg per tree at the time of full bloom would alleviate the disorder. Balanced fertilizer application of 3-4 calcium sprays at 1 or 2 week interval before harvest or post harvest dips in $CaCl_2$ solution with diphenylamine will also help to alleviate the problem. Controlled atmospheric storage and waxing can control bitter pit.

3. Apple scald

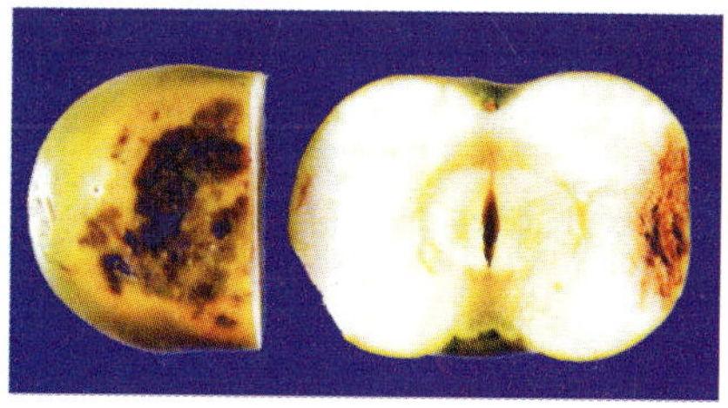

This is characterised by softening of fruit tissue with brown spots in the flesh during warm weather and water stress.

Control

This disorder can be controlled by coating fruits with paraffin wax, Vaseline or olive oil. Use of diphenylamine (10 ppm) and ethoxyquin (3ppm) will effectively prevent scald development, when applied with in a week after harvest. Fatty acid methyl esters and edible fats and oils can also control apple scald

9. SAPOTA

Fruit drop

Sapota has the problem of low fruit setting and shedding. Only about 10-12 percent of the total fruits set, develops and retained until maturity. Most of the fruit-drop occurs immediately after fruit setting. Increase in fruit set and retention are possible by spraying NAA and GA_3 at 25 to 100 ppm during flowering and at 15-day interval.

10. POMEGRANATE

Fruit cracking

Fruit cracking is a serious problem of pomegranate. The malady is thought to be due to boron deficiency in young fruits while in developed fruits it may be caused due to variations in soil moisture content and atmospheric humidity. At the time of fruit ripening, if the soils become too dry and then irrigated heavily or there is some rains, cracking may occur. Some cultivars, like Guleshan, Khog, Kazaki are reported to be resistant to fruit cracking. Regular irrigation to maintain soil moisture at desired level, spraying of calcium compounds or GA_3 at 120 ppm on young fruits are reported to mini-mize the fruit cracking.

Other common physiological disorders of fruit crops

Fruit skin splitting and cracking

These are serious physiological disorders causing substantial economic losses in several fruit crops. The cracks and splits appear on the fruit skin while the fruit is attached to plant and are mostly observed in pome and stone fruits. Cracking and splitting of detached fruit also occurs during post harvest handling and cold storage. The cracks and splits provide open wounds facilitating rapid moisture loss, excessive shriveling and, thus permitting infection by fungi. Fractures in the cuticle or skin may be microscopic or easily seen, extending deep in the flesh as well-defined cavities. The disorder is characterised by physical rupture of the fruit skin, resulting from various stresses acting on fruit. These can be due to normal growth processes or damage-induced. Splitting is the extreme form of cracking, whereas cracks may penetrate deep into the flesh, causing gross exposure of internal tissue to the atmosphere.

Causes of skin splitting and cracking

Sudden increase in soil water content or atmospheric humidity or temperature, osmotic absorption of water through fruit skin, and unusual supply of water given suddenly after a prolonged dry period are the probable causes of this common disorder.

Remedy

Foliar spray of daminozide (2000 ppm) two months before harvest controls cracking and splitting. Inheritance of fruit cracking and splitting and biophysical attributes that govern resistance and susceptibility of fruits to

5

REMEDIAL MEASURES

NITROGEN

The deficiency of nitrogen may be rectified by applying required quantity of nitrogen to soil or spraying on leaves (foliar application). Any one form of nitrogen may be applied to soil as top dressing followed by copious irrigation to rectify the deficiency. For foliar application, urea is used as a source of nitrogen. For a better growth and to avoid deficiency symptoms about 2 to 4% of urea solution may be sprayed depending upon the type of crop. In commercial urea, a toxic substance called biuret is present which will scorch the foliage if the quantity exceeds 2.0 per cent. The upper layer of urea solution carries biuret which could be easily removed before applying to the foliage. In the urea solution a scum like float is seen on the surface containing maximum quantity of biuret. The same has to be removed to avoid the scorching effect. Spraying is preferable during early hours of morning before 9.00 a.m. or in the evening around 4.00 p.m.

PHOSPHORUS

The requirement of phosphorus varies with the type of crop and nature of soil. Usually phosphorus is applied either before or during sowing/planting time. The type of phosphorus to be chosen for use is largely based on the soil reaction, relative costs and ease of application. For soil of neutral or alkaline condition, phosphatic fertilizers having high water soluble phosphorus are preferable. On the other hand such type of phosphorus fertilizer is not insisted for acid soils. Thus, the phosphorus deficiency in crop plants can be removed by adding the required quantity of the right type of phosphorus bearing in mind the factors involved in the soil and the crop.

In recent years spraying phosphate in the form of diammonium phosphate (DAP) at 2 per cent concentration to pulses has been proved to be good in increasing the yield of the crop. Application of DAP to rice nursery helps in producing robust and vigorous seedlings with better root growth. Attempts have been made to incorporate superphosphate with compost in the making of super digest compost, which proved to be more useful in improving the growth and yield of crop plants.

POTASSIUM

Application of organic mulches like farmyard manure or wheat straw or soybean hay to potassium deficient peaches growing on a sandy soil, corrected the deficiency symptoms, whereas the untreated trees on the same soil showed deficiency symptoms. Further experimental evidences show that application of potassium chloride along with the above organic amendments resulted in the effective penetration of applied potassium in soil which are found to convert potassium in to non- available form.

Potassium requirement of plant may also be met with either by applying the required quantity of potash fertilizer to soil or by spraying an optimum dose of potassium salt on the foliage. Muriate of potash is the common potash fertilizer used for soil application, which contains 60 per cent potassium. According to the availability in the soil and requirement of crop the quantity will be applied invariably as basal dressing for the crop. In case of spraying, excess dose may scorch the leaves and therefore special care should be taken with regard to the concentration of spray fluid.

Salt like potassium sulphate is commonly used for foliar spray. The tolerance towards the strength of potassium may vary from crop to crop. Sprays containing adequate but noninjurious amount of potassium salt is most effective. A concentration of one per cent has been tried for apple and as high as 7-10 per cent for potato.

MAGNESIUM

Magnesium deficiency in plants can be corrected by using epsom salt ($MgSO_4$ $7H_2O$) as soil application or foliar spray. It contains 9.6 per cent Mg. But foliar spray acts quickly and cures the malady within a fortnight, while soil application requires longer time to bring forth such results. In general, two per cent epsom salt solution proves to be highly efficient in correcting the maladies of deficiency.

For correcting the magnesium deficiency of cotton, a foliar spray of 0.5% $MgSO_4$ with 1.0 per cent urea and 0.1% zinc sulphate (500 g MgSO4, 1.0 kg urea and 100 g zinc sulphate in 100 litres of water) at 55th and 75th day was found to be effective. Depending upon, the severity of the deficiency, the strength of $MgSO_4$ and frequency of application may be increased if necessary.

Dolomitic lime stone ($MgCO_3$ + $CaCO_3$) can also be applied to soil but only under certain conditions of acidic to neutral range of soils and to non calcareous soils.

CALCIUM

Usually calcium is supplemented in the soil by the addition of dolomitic limestone, rock phosphate, burnt lime, hydrated lime, gypsum etc., depending upon the nature of the soil and the purpose.

In case of groundnut crop, gypsum is added at the rate of 80-100 kg/acre during the peg (gynophores) formation on the 45-55th day after sowing. This has been a regular schedule of practice with the farmers to apply gypsum to groundnut crop for increasing the yield and filling up of pods.

IRON

The corrective measure for iron deficiency has to be adopted according to the cause of deficiency. Addition of iron salts to the soil will prove to be a positive approach if the deficiency is owing to the lack of the element in soil, but the same measure may be unsuccessful if the deficiency is because of some restraining factors that influence the iron availability. Under such unfavourable conditions, the salt could be made easily available by applying the same in chelated form. Chelating agents are chemical substances, which are naturally present in soil. They are, humic acid, citric acid, tartaric acid, ascorbic acid and amino acid. These substances have got the ability to convert certain chemicals in to a soluble and available form. There are some synthetic chelating agents, which are identified for agricultural purposes.

They are Ethylene diamine tetra acetic acid (EDTA). Hydroxy ethyl diamine acetic acid, Diethylene triamine pentaacetic acid, Ethylene diamine di-o-hydroxy-phenylacetic acid and Amino polycarboxylic acid.

The acid molecules of the deleting agents chelate the metal ions in two or more positions within the molecular structure. In case of iron chelate, the Fe ion is bound and it cannot release itself to form another compound even when treated with phosphate or hydroxide. The iron chelate is quite soluble in water and becomes available to plants as a nutrient iron. Both the iron and chelating molecule may enter the plant together and transported to the leaves without undergoing any undesirable fixation of the molecule.

A good chelating agent must resist hydrolysis, microbial break-down, fixation by clay particles, but must be ready to release the element to the plant when required and should not be toxic to plant either when applied to foliage or soil. Further it should not leave any toxic nature of residue in edible portion of plants.

Soil application of iron chelate is more beneficial and successful than foliar application. A quantity of 11.25-22.5 kg/ac of iron chelate is applied for field crops and a quantity of 1 to 2.25 kg/tree is applied for fruit trees.

Foliar spraying is done with ferrous sulphate at a concentration of 2-4 per cent depending upon the crop and severity of deficiency. Iron deficiency found in sugarcane is being corrected successfully by applying ferrous sulphate at 2.5 kg in 450 litres of water with an addition of 1.25 kg lime.

MANGANESE

Manganese deficiency may be corrected by application of $MnSO_4$ to soil. The quantity to be applied to soil will vary according to the type of soil. Large quantities of 75-150 kg of $MnSO_4$ / acre may have to be applied to problem soils (alkaline soils). This involves very high expenditure.

Manganese can also be applied to plants as foliar sprays with manganese sulphate. The quantity required in this method would be small. The expenditure on account of it will be very little and therefore it is economical. Further the recovery is quick since the element directly enters into the levels through stomata. Foliar application proves to be more successful and hence it is more popular.

Manganese deficiency of citrus can be corrected by spraying manganese sulphate at 0.2-0.4 per cent. The solution should be neutralized with lime (half the quantity of salt). Although the injection of manganous sulphate to the citrus, peach and cherry trees rectified the deficiency, foliar spray has been more popular in adoption.

ZINC

Correction of zinc deficiency is attempted by application of zinc salts like sulphate through soil or foliar spray or seedling dip. In fruit trees injection of zinc salt or driving 20-30 zinc coated nails in trunks of big trees were also followed. But the degree of success varies in each method of application depending upon the crop, soil and climate. In general, soil application and foliar spray are found to be more successful. As foliar spray, a dose of 0.2-0.5 per cent zinc sulphate with half the quantity of lime will effectively correct the deficiency. The quantity of zinc for soil application will vary depending on the crop and soil.

Dipping the rice seedlings in 2.0 per cent zinc sulphate solution for 5-10 minutes before planting will prevent the deficiency in planted crop. In the transplanted crop, application of 12 kg of zinc sulphate once in two years in alkaline soils will improve the growth of the crop. It would also enhance the performance of the crop when the land is applied heavily with farmyard manure or green leaf manure. Foliar spray with 0.2 per cent zinc sulphate can also be done under certain conditions.

For citrus, foliar spray with 0.5 per cent of zinc sulphate neutralized with lime to half the quantity of the former will correct the deficiency.

COPPER

Copper salt like basic copper sulphate can be applied to soil at the rate of 10 kg / acre or be sprayed at one per cent level depending upon the necessity and nature of the crop. The latter method will be more desirable than soil application. Foliar application of copper is usually adopted in many crops as a measure to check fungal diseases. While resorting to foliar application, neutralization of spray fluid with equal quantity of lime is quite necessary.

BORON

The deficiency of boron can be corrected by application of boron either by foliar spray or through soil. For foliar sprays boron compounds like borax or boric acid may be used but for soil, borax is commonly used and it usually remains effective for longer periods. In western countries several fertilizers are available with boron as an ingredient in minute quantities. Such fertilizers should be applied just before sowing or after sprouting. It should not be mixed with seed while sowing since it will reduce the germination of seeds.

Factors affecting nutrient uptake

There are numerous factors that affect uptake of fertilizers. The most important of these is fertilizer form. Generally, inorganic, fast-release, or liquid forms are faster-acting than organic, slow-release, or dry forms. Compared to synthetic fertilizer formulations, most organic fertilizers contain relatively low concentrations of actual nutrients.

- **Soil composition:** Soils high in clay particles will absorb (bind to their surface) more nutrients, while fertilizer will leach (wash through) faster through sandy soil. Organic matter in the soil increases its nutrient holding capacity and contributes nutrients upon its breakdown.
- **Soil microorganisms:** Some fungi and bacteria may "tie up" nutrients while others convert the fertilizer to a form that the plant can take up. Some microorganisms are involved in mutually beneficial (symbiotic) relationships with plants. Rhizobium bacteria grow within the roots of some plants. They convert nitrogen from the air into a usable form for the plant while obtaining nutrition from the host plant.
- **Soil pH:** Extremes in pH affect availability of plant nutrients and the concentration of plant-toxic minerals. At low pH levels, calcium, phosphorus, and magnesium become unavailable and manganese can concentrate at toxic levels. At high pH values, phosphorus, iron, copper, zinc, boron, and manganese become less available.
- **Nutrient availability:** Nutrients may be present but may first require conversion to an "available" form that the plant is capable of taking up and utilizing. Conversion to an available form is affected by soil microorganisms, pH, soil moisture and chemical reactions.
- **Soil moisture content:** Most nutrients are taken up via the soil solution, so soil water is needed to dissolve them.
- **Soil aeration:** Oxygen is needed in the soil to help roots with uptake processes. Where there is no oxygen, such as in flooded sites, sugar cannot be utilized by the plants to produce energy for nutrient uptake. Decomposed organic matter helps develop good air-water relationships in the soil.
- **Soil temperature:** Nutrient uptake is faster in warmer soils than in cold soils.
- **Plant condition:** Plants under stress will be less able to take up nutrients, generally due to a reduced or damaged root system.
- **Competition:** If the roots of many plants occupy an area, a reduced amount of nutrients will be

available for each. When using close spacing for vegetable plants, more fertilizer will be needed in a bed than in a conventionally spaced row garden. Weeds where present will take up nutrients intended for landscape or crop plants. Reduction of weeds will reduce fertilizer needs.

6

FOLIAR NUTRITION

This refers to the application of a nutrient on the leaves. It requires a sprayer and some knowledge regarding the preparation of the spray solution correctly. With ordinary high volume sprayers, usually 400 litres of spray solution are needed per hectare. Some typical concentrations of spray solution are given below:

Mg : 1% magnesium sulphate

B : 0.2 - 0.5% soluber

Cu : 0.025% (100 g Cu as $CuSO_4$) 3 0 g EDTA / ha

Fe : 1 - 3 per cent ferrous sulphate

Mn : 0.4-2% manganese sulphate

Mo : 0.1-0.3% ammonium molybdate

Zn : 0.5% zinc sulphate, 0.25% lime

These sprays have usually to be repeated 2, 3 or more times at 7- 15 day intervals in order to bring the deficiency under complete control. Lime is usually added, where recommended, to obtain a near neutral solution which should be filtered in order to use only a clear solution for spray. The relative features of a soil application versus foliar application are given below.

Advantages of foliar application	Advantages of soil application
Lower application rates needed	Enables application of deficient nutrient at the right time
Uniform application of small amounts	Provides residual benefit to succeeding crop
Immediate effect on the crop; No build up to harmful levels	No equipment (sprayer etc.) needed
Good supplement to soil application as a mid course correction	Application conditions flexible
Suspected deficiencies can be easily confirmed	Skilled labour less important
Main practical method for tree crops	Nutrient application can be integrated with usual basal dose

1 kg $ZnSO_4$ $7H_2O$ = 210 g Zn

1 kg borax $Na_2B_4O_7.10H_2O$

= 365 g B_2O_3 = 113 g B

The following method can be used to calculate the amount of a fertilizer required to provide the recommended amount of a nutrient.

kg fertilizer needed =

$$\frac{\text{kg nutrient needed}}{\text{\% nutrient in fertilizer}} \times 100$$

For example:

kg ammonium sulphate needed to get 5 kg S=5/24x100=20.8 kg sulphate

kg zinc sulphate needed to get 10 kg zinc = 10/21 x 100 = 47.6 kg

kg borax needed to get 1 kg B = 1/11.4 x 100 = 8.8 kg Borax

These are types of calculations with which every extension worker, salesman, dealer, or progressive farmer should be familiar. As a ready information, the following can be used.

To obtain	kg of the material needed
1 kg Ca	2.5 kg limestone or 5 kg dolomite
1 kg Mg	10.4 kg $MgSO_4$ $7H_2O$ or 6.2 kg . $MgSO_4$. H_2O or 8 kg dolomite
1 kg S	4.2 kg AS, or 8.3 kg SSP or 6.7 kg APS
1 kg B	8.8 kg borax or 4.8 kg SOLUBER
1 kg Cu	4.2 kg copper sulphate
1 kg Fe	5.3 kg ferrous sulphate or 8.3 kg Fe EDTA
1 kg Mn	3.3 kg manganese sulphate
1 kg Mo	1.92 kg ammonium molybdate
1 kg Zn	4.8 kg $ZnSO_4$7 H_2O or 8.3 kg Zn EDTA

A ready reckoner for calculating fertilizer quantities to obtain the required amount of an element or oxide is given in table. As an example, if 5 kg Zn is to be applied through zinc sulphate then against 5 in column 1, may go horizontally till one reaches the column heading of 20 (because zinc sulphate has about 20 % Zn). The value of 25 in this column gives the quantity of zinc sulphate to be applied to provide 5 kg Zn. The more concentrated a fertilizer, smaller would be the quantity needed to apply a given amount of the nutrient.

Quantity of fertilizer material needed to obtain a given quantity of the nutrient

Kg nutrient needed element or oxide	Kg Fertilizer needed (% nutrient or oxide in fertilizer)					
	5	10	20	30	40	50
0.5	10	5	2.5	1.7	1.25	1
1	20	10	5	3.3	2.5	2
2	40	20	10	6.7	5	4
3	60	30	15	20	7.5	6
4	100	50	25	16.7	12.5	10
5	200	100	50	33	25	20

For spray solutions, fertilizer needed to obtain the desired concentration is given below, assuming that 400 litres solution is required per hectare. This shows that at 0.5 per cent concentration one needs 2 kg fertilizer per ha/spray.

The usually advocated spray concentration with sulphate salts is 0.5-1.0 per cent but can vary quite a bit depending on the crop and local conditions.